W0017991

Assessment of the Risk of Amazon Dieback

Walter Vergara and Sebastian M. Scholz, editors

THE WORLD BANK
Washington, D.C.

Library of Congress Cataloging-in-Publication Data

Vergara, Walter, 1950-
 Assessment of the risk of Amazon dieback / Walter Vergara, Sebastian M. Scholz.
   p. cm. -- (World Bank study)
 Includes bibliographical references and indexes.
 ISBN 978-0-8213-8621-7 -- ISBN 978-0-8213-8622-4 (electronic)
 1. Forest microclimatology--Amazon River Region--Computer simulation. 2. Climatic changes --Amazon River Region--Forecasting--Computer simulation. 3. Forest biomass--Carbon content-- Amazon River Region--Computer simulation. 4. Rain forest plants--Climatic factors--Amazon River Region--Computer simulation. 5. Deforestation--Amazon River Region--Computer simulation.
 I. Scholz, Sebastian M. II. Title. III. Series: World Bank study.
   SD390.6.A48V47 2010
   333.75'1409811--dc22   2010039311

Contents

Tables

Figures

Preface

The Amazon basin is a key component of the global carbon cycle, which is itself a determining factor for global climate. The rainforests in the basin store about 120 billion metric tons of carbon in their biomass. Indeed, the Amazon rainforest is considered to be a net carbon sink or reservoir because vegetation growth on average exceeds mortality, resulting in an annual net sink of between 0.8 to 1.1 billion metric tons of carbon.

In Brazil, the basin is the home of about 25 million people, most in urban areas, but it also includes several unique indigenous and traditional cultures, and is the largest repository of global biodiversity. It is larger than the European Union (around 5.2 million square kilometers) and produces about 20 percent of the world's flow of fresh water into the oceans.

Current climate trends may be unbalancing this well-regulated system and, in association with land use changes, may be shifting the region from a carbon sink to a carbon source. Changing forest structure and behavior would have significant implications for the local, regional and global carbon and water cycles. Amazon forest dieback would be a massive event, affecting all life-forms that rely on this diverse ecosystem, including humans, and producing ramifications for the entire planet. Clearly, with changes at a global scale at stake, there is a need to better understand the risk, and dynamics of Amazon dieback.

Thus, the goal of this study is to assist in understanding the risk of a potential reduction in biomass density in the Amazon basin induced by climate change impacts (Amazon dieback) and its implications.

Feedback and fertilization effects of elevated atmospheric carbon dioxide levels on forest ecosystems like the Amazon have proven to be a key unknown in assessing the risk of Amazon forest dieback under 21st century climate change scenarios. Reducing this uncertainty ought to be a key priority going forward. In the absence of robust information, the precautionary principle applies, which in this case suggests that the assumption of carbon dioxide fertilization being an important factor positively affecting ecosystem resilience of the Amazon cannot be used as a basis for sound policy advice.

Therefore, the study concludes that during this century, the probability of Amazon dieback is highest in the Eastern Amazon and lowest in the Northwest, but that its severity increases over time and also is a function of the global greenhouse gas emission trajectory considered. These results point to the need to avoid reaching a point in global emissions that would result in an induced loss of Amazon forests. Therefore, Amazon dieback should be considered a threshold for dangerous climate change.

Walter Vergara
Leader
Global Expert Team on Climate Change Adaptation
The World Bank

Sebastian M. Scholz
Environmental Economist
Sustainable Development Department,
Latin America and Caribbean Region,
The World Bank

Acknowledgments

This study is a joint effort of the World Bank (team led by W. Vergara with the participation of S. M. Scholz, A. Deeb, N. Toba, A. Valencia, A. Zarzar, and K. Ashida) with the collaboration of several institutions, namely the Meteorological Research Institute of Japan (team led by A. Kitoh), Exeter University of England (team led by P. Cox and T. Jupp), the Potsdam Institute for Climate Impact Research (team led by W. Lucht, W. Kramer and A. Rammig), CCST/INPE Brazil (team led by C. Nobre and G. Sampaio and participation among others of M. Cardoso, L.F. Salazar, and J. Marengo), and GEXSI of Germany (team led by M. Koch-Weser). The full list of authors can be found in Appendix 4.

The results of the analysis were reviewed by a blue-ribbon panel of internationally renowned scientists and practitioners. The panel members were Thomas E. Lovejoy (Chair), The H. John Heinz Center for Science, Economics, and the Environment; Lawrence E. Buja, National Center for Atmospheric Research; David Lawrence, National Center for Atmospheric Research; Michael T. Coe, University of Wisconsin-Madison; Earl Saxon, Independent Consultant; and Ben Braga, University of São Paulo.

The team is also most grateful to M. Diop, J. Nash, E. Fernandez, M. Lundell, L. Tlaiye, D. van der Mensbrugghe, G. Batmanian, T. Cordella, and C. de Gouvello for their comments and suggestions and to I. Leino for her help in editing the document.

Finally, acknowledgements are due to the team of Brazilian Government representatives (led by Artur Cardozo de Lazerda, from the Ministry of Finance) that provided insightful comments and suggestions to clarify and strengthen the presentation of the results.

The resulting report was edited by W. Vergara and S. Scholz on the basis of the technical inputs produced under contract from the different teams above indicated.

This study is funded through the Climate Change and Clean Energy Initiative, and is a product of the Environment Unit of the Sustainable Development Department of the Latin America and Caribbean Region of the World Bank. The team is most grateful to the Country Management Unit of Brazil and the Global Expert Team on Climate Change Adaptation at the World Bank for their help in publishing this study.

Acronyms and Abbreviations

ANSG	Atlantic North-South Gradient
AGCM	Atmospheric General Circulation Model
AOGCM	Atmospheric Ocean General Circulation Model
AR4	Fourth Assessment Report of Intergovernmental Panel on Climate Change
CDD	Consecutive Dry Days
CDF	Cumulative Distribution Function
CGCM	Coupled General Circulation Model
CSIRO	Australia's Commonwealth Scientific and Industrial Research Organisation
CCST	Center for Climate Science and Technology
CMIP3	Coupled Model Intercomparison Project 3
CPTEC	Centro de Previsão de Tempo e Estudos Climáticos
CRU	Climatic Research Unit–University of East Anglia
DGVM	Dynamic Global Vegetation Model
DJF	December January February (Wet Season)
ENSO	El Niño Southern Oscillation
FC	Forest Cover
FCP	Forest Conservation Plan
FPC	Foliar Projective Cover
GCM	General Circulation Model
GHG	Greenhouse Gases
GPCP	Global Precipitation Climatology Project
GRDC	Global River Discharge Center
IBIS	Integrated Biosphere Simulator Model
INPE	Brazilian Institute for Space Research
IPCC	Intergovernmental Panel on Climate Change
ITCZ	Intertropical Convergence Zone
JJA	June July August (Dry Season)
JMA	Japan Meteorological Agency
LPJmL	Lund-Potsdam-Jena managed Land Dynamic Global Vegetation and Water Balance Model
MAM	March April May (Wet Season)
MRI	Meteorological Research Institute
NPP	Net Primary Production
ORCHIDEE	Organizing Carbon and Hydrology In Dynamic Ecosystems
PDF	Probability Density Function
PEWG	Pacific East-West Gradient

PFT   Plant Functional Type
PgC   Pentagram Carbon
RX5D   Maximum 5-day precipitation total
SACZ   South Atlantic Convergence Zone
SON   September October November (Dry Season)
SRES   Special Report on Emission Scenarios
SST   Sea Surface Temperature
TDO   Task Development Objective

CHAPTER 1

Introduction

The Amazon basin is a key component of the global carbon cycle. The old-growth rainforests in the basin represent storage of ~ 120 petagrams of carbon (Pg C, equal to 120 billion metric tons of carbon) in their biomass. Annually, these tropical forests process approximately 18 Pg C through respiration and photosynthesis. This is more than twice the rate of global anthropogenic fossil fuel emissions (Dirzo and Raven 2003). The basin is also the largest global repository of biodiversity and produces about 20 percent of the world's flow of fresh water into the oceans. Despite the large CO_2 efflux from recent deforestation, the Amazon rainforest ecosystem is still considered to be a net carbon sink of 0.8–1.1 Pg C per year because growth on average exceeds mortality (Phillips et al. 2008).

However, current climate trends and human-induced deforestation may be transforming forest structure and behavior (Phillips et al. 2009). Increasing temperatures may accelerate respiration rates and thus carbon emissions from soils (Malhi and Grace 2000). High probabilities for modification in rainfall patterns (Malhi et al. 2008) and prolonged drought stress may lead to reductions in biomass density. Resulting changes in evapotranspiration and therefore convective precipitation could further accelerate drought conditions and destabilize the tropical ecosystem as a whole, causing a reduction in its biomass carrying capacity or dieback. In turn, changes in the structure of the Amazon and its associated water cycle would have implications for the many endemic species it contains and result in changes at a continental scale. Clearly, with much at stake, if climate-induced damage alters the state of the Amazon ecosystem, there is a need to better understand its risk, process, and dynamics.

Objective

The objective of this study is to assist in understanding the risk, process, and dynamics of potential Amazon dieback and its implications. The task is organized in five activities that address key aspects of the analysis (see figure 1.1): a) a modeling of future climate (end of century) over the basin, using a high-resolution model; b) an assessment of the impact of climate on rainfall over the region; c) an assessment of biomass response to rainfall anomalies and associated changes; d) an assessment of linkages between deforestation and potential for dieback; and e) on the basis of these assessments, a qualification of potential economic consequences is described.

Scope

Amazon dieback is defined as the process by which the Amazon basin loses biomass density as a consequence of changes in climate. Although there is no consensus definition on

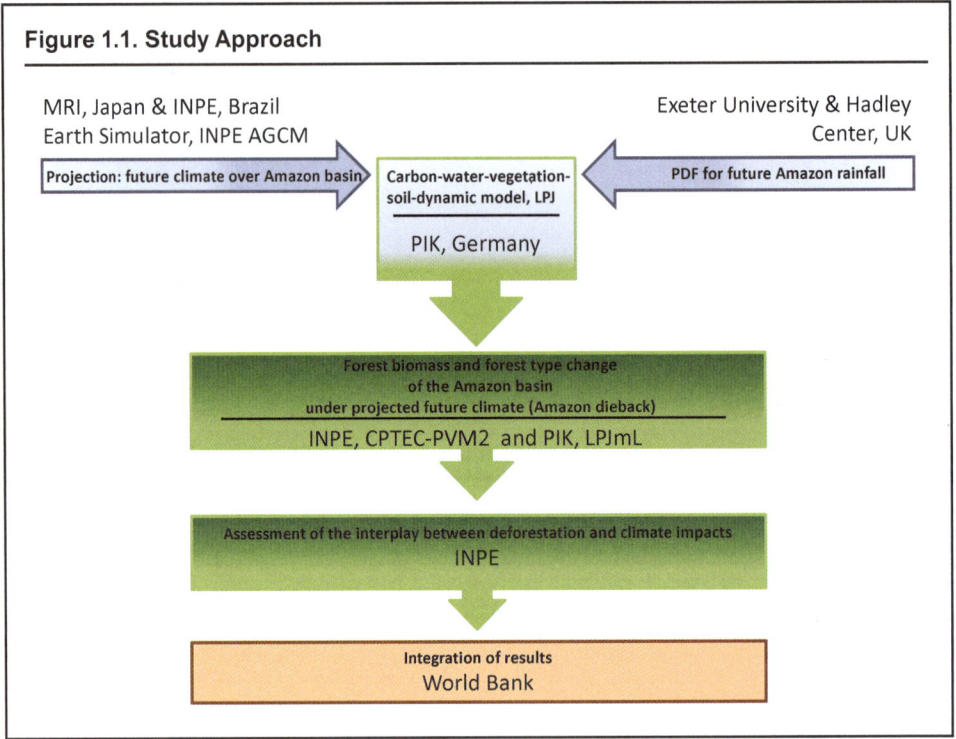

Figure 1.1. Study Approach

Source: Figure generated for the report.

how to characterize forest dieback, for purposes of this analysis reductions in biomass carbon resulting from climate impacts, which would exceed 25 percent of the standing stock of carbon, are considered to be an indication of dieback.

The analysis considers that climate-induced biomass response in the basin will result not only from changes in rainfall but also from other factors that are linked to climate changes, such as: a) increased probabilities for prolonged drought stress, which may lead to increased physiological stress for trees, increased tree mortality, and thus carbon emissions; b) increasing atmospheric CO_2 concentrations, which may alter the drought response of forests; c) increasing temperatures, which may accelerate heterotrophic respiration rates and thus carbon emissions; and d) resulting changes in evapotranspiration and therefore convective precipitation, which could further accelerate drought conditions and destabilize the tropical ecosystem as a whole.

The structure of the analysis is presented in figure 1.1. Results from the modeling of future climate in the Amazon basin, based on the high-resolution atmospheric general circulation model of the Meteorological Research Institute of the Japan Meteorological Agency, and using the Earth Simulator, are reported in chapter 2 (**Modeling Future Climate in the Amazon Using the Earth Simulator**). It summarizes estimates of future climate extremes (using indexes for extreme wet and extreme dry periods), rainfall, soil dryness, runoffs, and stream flows, under different greenhouse gas emission trajectories or Special Report on Emission Scenarios (SRES) (see an explanation of SRES scenarios in annex 1). The outputs of the model are used to estimate biomass response in chapters

4 and 5. Chapter 3 (**Assessment of Future Rainfall over the Amazon Basin**) examines the difficulties of predicting future rainfall over the region and reviews the outputs from 24 Global Circulation Models (GCM) used by the Intergovernmental Panel on Climate Change (IPCC) from a perspective of the ability to simulate current rainfall, akin to an "Amazon Prediction" index. Chapter 3 also presents probability density functions of future rainfall based on this index and on the ability to reproduce sea surface temperature dipoles, for use in assessing biomass response.

Chapter 4 (**Analysis of Amazon Forest Response to Climate Change**) presents the results of the application of the Lund-Potsdam-Jena managed Land Dynamic Global Vegetation and Water Balance Model (LPJmL) to the outputs and the implications of rainfall changes. It results in estimates of future vegetation carbon in the region in the form of probability density functions. Finally, chapter 5 (**Interplay of Climate Impacts and Deforestation in the Amazon**) presents the potential combined effects of climate change and deforestation in the Amazon, using the results of the Earth Simulator, the outputs of the Center for Weather Forecasting and Climate Studies Global Circulation Models (*Centro de Previsão de Tempo e Estudios Climáticos*, CPTEC GCM), and the CPTEC vegetation model. Chapter 6 (**Conclusions**) summarizes the impacts that would be expected.

Geographical Domain

Because the Amazon basin covers a wide region subject to different stresses and conditions, all assessments have focused on five geographical domains (windows) defined to capture different momentums in land use. Eastern Amazonia is a region continuous to Northeastern Brazil, where somewhat drier conditions are already the norm and growing anthropogenic impacts can be observed. Northwestern Amazonia is a region with little if any current direct anthropogenic impact and relatively intact ground cover. Southern Amazonia is a region subjected to strong land use change drivers. Northeastern Brazil is a region subjected to dry conditions. Southern Brazil[1] would suffer the consequences of any change in climate in Amazonia (see figure 1.2).

Data Sources

Observations: The Climate Research Unit (CRU) Dataset

Observed data for air temperature (Celsius, °C), precipitation (millimeters, mm), cloud cover (percent) and the number of wet days at a 0.5° resolution (latitude/longitude grid) were available for monthly time-steps throughout the 1901–2003 period from the Climate Research Unit (CRU), University of East Anglia (New et al. 2000). These data are assumed to represent the "true" state of the climate (from the CRU TS 3.0 archive: http://www.cru.uea.ac.uk/cru/data/hrg-interim/). Three additional climatology archives were used for comparison purposes: the CMAP, GPCP, and TRMM 3A25[2] datasets.

Greenhouse Gas Emission Trajectories (SRES scenarios)

The projected global greenhouse gas emission trajectories (SRES scenarios) adopted by the IPCC and used in its fourth assessment report (AR4), cover a wide range of driving forces of future emissions, including demographic, technological, and economic devel-

Figure 1.2. Geographical Domains

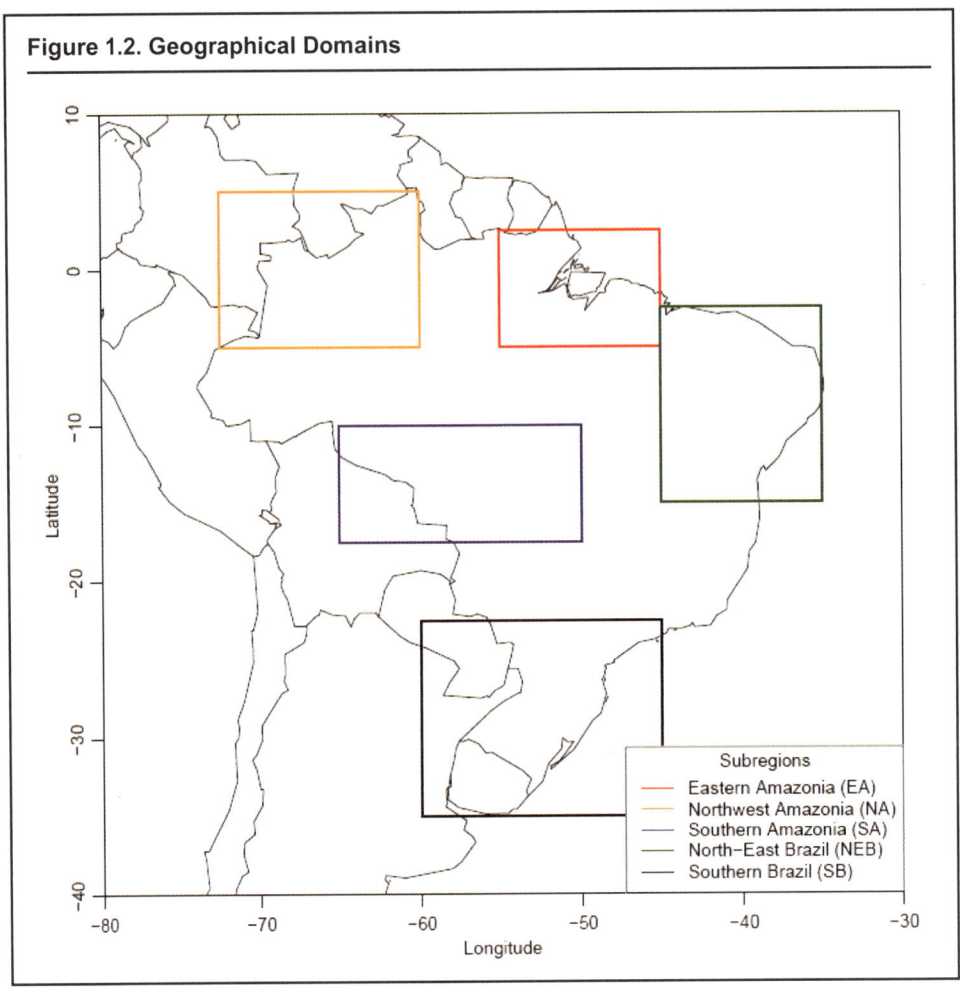

Source: Figure generated for the report by Cox and Jupp 2009.
Note: Eastern Amazonia (EA; 2.5°N-5°S; 45°W-55°W); Northwestern Amazonia (NA; 5°N-5°S; 60°W-72.5° W); Southern Amazonia (SAz; 10°S-17.5°S; 50°W-65°W; Northeastern Brazil (NEB; 2.5°S-15°S; 35°W-45° W; Southern Brazil (SB; 22.5°S-35°S; 45°W-60°W).

opments. The Amazon dieback study is based on the scenario of moderate but consistent improvements in energy efficiency and deployment of renewable energy (called A1B), which estimates an end-of-century temperature anomaly (increase) of 2.8 degrees Celsius. This has been used until recently as the mid-range scenario for climate work under the IPCC. However, this scenario may be an underestimate, as the current trajectory already far surpasses the estimates used to define it. For the analysis of the interplay between deforestation and climate change, the Amazon Dieback report uses two more scenarios, one where fossil fuels remain predominant (A2) and one with a greater penetration of renewable and more significant gains in energy efficiency (B1), with respective net increases in temperature of 3.4 to 1.8 degrees Celsius respectively, that provide a wider set of estimates than A1B.

IPCC AR4 Coupled General Circulation Models

Climate outputs were available from the Coupled General Circulation Models (CGC-Ms) that participated in the Coupled Model Intercomparison Project 3 (CMIP3; coupled atmosphere-ocean models) carried out for the IPCC's Fourth Assessment Report (IPCC 2007). These data comprise the output from 24 climate models (available at https://esg.llnl.gov:8443/).

High-Resolution GCM Data

In addition, super-high-resolution (20 km) and high-resolution (60 km) temperature, precipitation, and cloud cover simulations from the Meteorological Research Institute (MRI) and Japan Meteorological Agency (JMA) Atmospheric Global Circulation Model (AGCM) using the Earth Simulator were available for three periods: The 1979–2003 period represents the present climate conditions (P conditions). Scenario output for future conditions under the SRES-A1B emission scenario is available for the 2015–2039 period (near future, N conditions) and the 2075–2099 period (far future, F conditions).

Sea Surface Temperature (SST)

Four different SSTs are used for future climate simulations by the 60-km mesh model. One experiment uses the CMIP3 model ensemble SST and sea-ice distributions as in the 20-km mesh model experiment. Second, third, and fourth experiments use the SST anomalies of Australia's Commonwealth Scientific and Industrial Research Organisation (CSIRO)-Mk3.0, MRI-CGCM2.3.2 and MIROC3.2 (hires) models.

Table 1.1. Projected Global Average Surface Warming and Sea Level Rise at the End of the 21st Century According to Different SRES Scenarios

Case	Temperature change (degrees centigrade at 2090-2099 relative to 1980-1999) [a,d]		Sea level rise (meters at 2090-2099 relative to 1980-1999
	Best estimate	*Likely* range	Model-based range (excluding future rapid dynamical changes in ice flow)
Constant year 2000 concentrations [b]	0.6	0.3-0.9	Not available
B1 scenario	1.8	1.1-2.9	0.18-0.38
A1T scenario	2.4	1.4-3.8	0.20-0.45
B2 scenario	2.4	1.4-3.8	0.20-0.43
A1B scenario	2.8	1.7-4.4	0.21-0.48
A2 scenario	3.4	2.0-5.4	0.23-0.51
A1FI scenario	4.0	2.4-6.4	0.26-0.59

Source: IPCC 2007.

Notes: a) Temperatures are assessed best estimates and *likely* uncertainty ranges from a hierarchy of models of varying complexity as well as observational constraints.

b) Year 2000 constant composition is derived from Atmosphere-Ocean General Circulation Models (AOGCMs) only.

c) All scenarios above are six SRES marker scenarios. Approximate CO_2-eq concentrations corresponding to the computed radiative forcing due to anthropogenic GHGs and aerosols in 2100 (see p. 823 of the Working Group I TAR) for the SRES B1, AIT, B2, A 1 B, A2, and A1FI illustrative marker scenarios are about 600, 700, 800, 850, 1250, and 1550ppm, respectively.

d) Temperature changes are expressed as the difference from the period 1980-1999. To express the change relative to the period 1850-1899 add 0.5°C.

Vegetation Models

Two vegetation models were used to assess biomass response to various forcings. These are the LPJmL and the CPTEC-PVM. The LPJmL is a dynamic uncoupled model. The advantage of using LPJmL for biomass response to climate is that it is a process-based model that explicitly simulates the accumulation and loss of carbon, and vegetation dynamics. The CPTEC-CPVM is a static coupled model that simulates biome distribution (one biome per grid-cell) based on bioclimatic limits. The advantage of being coupled to a climate model is that feedbacks of vegetation change to the climate can be investigated. While LPJmL simulates biomass response, CPVM focuses on simulation of anticipated biome-equilibrium states. These two instruments complement each other. A summary of the inputs, processes, and outputs from each subtask is described in table 1.2.

Table 1.2. Summary of Inputs, Processes, and Outputs for Each Task

Task	Inputs	Emission trajectory	Process	Outputs
High-resolution simulation of future climate in the Amazon basin	MRI-GCM data (Earth Simulator)	A1B	High-resolution simulation to end of century though Earth Simulator	Future climate over the basin; projection of extreme events
Assessment of future rainfall over the Amazon basin	CMIP3 data (ensemble of 24 General Circulation Models)	A1B	Use of an Amazon climate prediction index to qualify CMIP3 outputs	Projected rainfall (weighted by model's ability to predict current climate) in the form of Probability Density Functions (PDF) of future rainfall
Forest biomass estimate	PDF results for future rainfall	A1B	LPJmL (dynamic uncoupled vegetation model)	Biomass response (weighted by rainfall prediction index); PDF for future biomass
Interplay of climate and deforestation	CMIP3	A2-B1	CPTEC-CPVM (static coupled vegetation model)	Biome shifts in the Amazon basin

Source: Table generated for the report.

Notes

1. The exact geographical coordinates are provided in figure 1.2.
2. These are the CPC Merged Analysis of Precipitation for 29 years (1979–2007) on a 2.5° lat/lon grid (CMAP: Xie and Arkin 1997), the GPCP One-Degree Daily Precipitation Data Set for 10 years (1998–2007) on a 1.0° lat/lon grid (GPCP: Huffman et al. 2001), and the Tropical Rainfall Measuring Mission (TRMM) PR3A25 V6 dataset for 9 years (1998–2006) on a 0.5° lat/lon grid (TRMM 3A25: Iguchi et al. 2000).

CHAPTER 2

Modeling Future Climate in the Amazon Using the Earth Simulator

The Atmospheric General Circulation Model Simulated by the Earth Simulator

As indicated in chapter 1, the Fourth Assessment Report of the Intergovernmental Panel on Climate Change (AR4) uses a dataset of 24 global coupled atmosphere-ocean general circulation models (AOGCM, or GCM for short) to project future climate under various scenarios. The use of numerous models is intended to reduce errors and uncertainty. However, most of these models have a very coarse resolution (100–400km) and this has an undesirable impact on results, particularly as it relates to extreme weather events. This is because global warming would result not only in changes in mean climate conditions but also in increases in the amplitude and frequency of extreme events that would not be captured in a meaningful way with coarse resolutions. Moreover, changes in extremes are more important for assessing adaptation strategies to climate changes. Therefore, a high spatial resolution model is required to study extreme weather events and to project their intensity and frequency for adaptation studies and measures.

The MRI/JMA atmospheric GCM is a super-high-resolution atmospheric general circulation model with a horizontal grid size of about 20 km (Mizuta et al. 2006), offering an unequaled high-resolution capability. The use of the supercomputer called the Earth Simulator made this super-high-resolution model's long-term simulation possible. The atmospheric GCM is a global hydrostatic atmospheric general circulation model developed by the MRI/JMA. This model is an operational short-term numerical weather prediction model of JMA and part of the next-generation climate models for long-term climate simulation at MRI. The data generated by the Earth Simulator was made available under the five-year Memorandum of Understanding between MRI and the World Bank. The outputs of the MRI/JMA GCM represent the anticipated changes in climatic conditions induced by the global emissions of greenhouse gases (GHG).

Although the global 20-km model is unique in terms of its horizontal resolution for global change studies with an integration period up to 25 years, available computing power is still insufficient to enable ensemble simulation experiments and this limits its application to a single member experiment. To address this limitation, parallel experiments with lower resolution versions of the same model (60-km, 120-km, and 180-km mesh) were performed. In particular, ensemble simulations with the 60-km resolution have been performed and compared with the 20-km version for this study.

The MRI-GCM was used to project climate in the Amazon basin to mid-century (2035–2049) and to the end of the 21st century (2075–2099) and compared these projections to the present (1979–2003) under scenario A1B,[1] which projects a temperature increase of between 1.7 and 4.4 degrees Celsius by the end of the century. The analysis was done primarily to assess rainfall, runoffs, and extreme events, and to estimate the anticipated impact on stream flows induced by climate change. Results on rainfall, moisture, and evaporation are also reported and later compared with other model outputs in subsequent chapters of the report. The simulations were performed at a grid size of about 20 km and routinely compared with 60-km mesh ensemble runs to ascertain robustness. A detailed description of the model and its performance in the 10-year present-day simulation with sea surface temperature (SST) can be found in Mizuta et al. (2006).

Comparison of Observed and Simulated Data for Present Time over the Amazon Basin

Seasonal mean precipitation reproduced in the simulation is evaluated against available observed data. Figure 2.1 shows the geographical distribution of December–February (DJF) averaged for a 25-year period (1979–2003) of mean precipitation for 180-km, 120-km, 60-km, and 20-km resolutions. Observations show large seasonal mean precipitation in the austral summer over the Amazon basin. The Intertropical Convergence Zone (ITCZ) over the tropical Atlantic, and the South Atlantic Convergence Zone (SACZ) to the southeast of the Brazilian Plateau, are also well reproduced. The precipitation maximum over the Amazon tends to locate northwest of observed data.

Figure 2.1. Geographical Distributions of December–February Mean Rainfall (mm d⁻¹) over the Amazon Basin

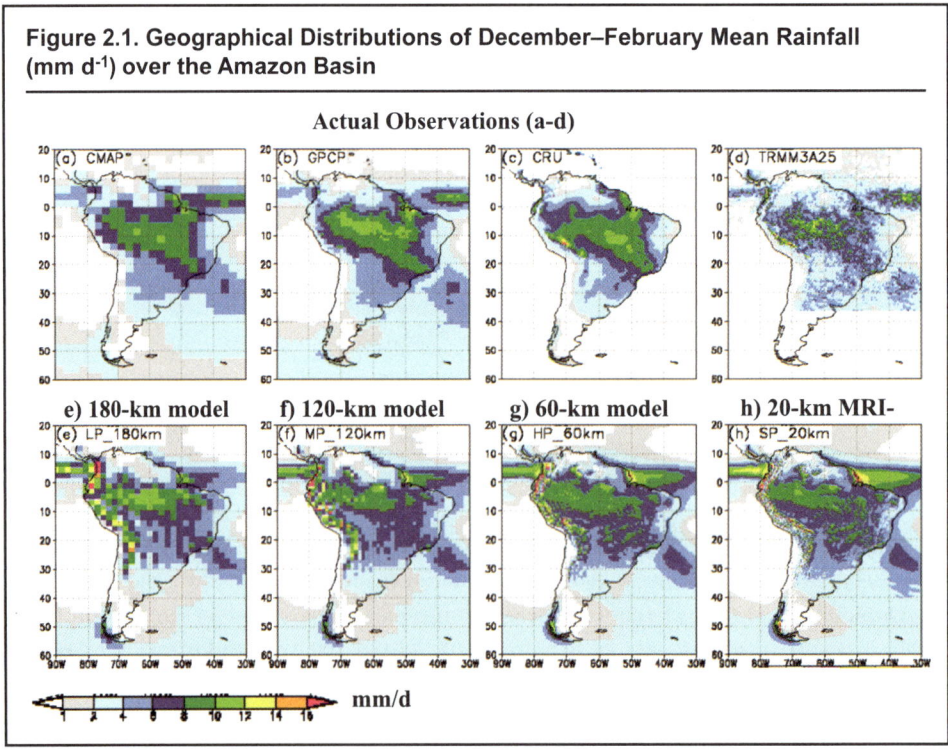

Source: Figure generated for the report by Kitoh et al. 2009.
Note: Plots (a-d) correspond to data-sets of actual observations; (e) is 180-km model, (f) 120-km model, (g) 60-km model and (h) 20-km MRI-GCM model

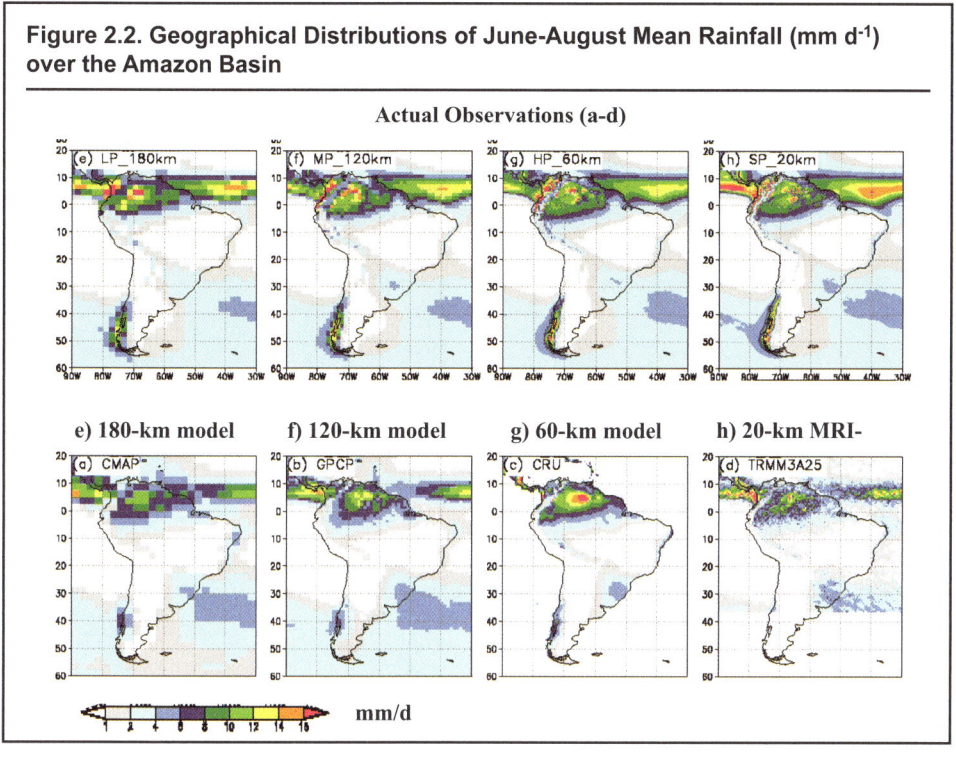

Figure 2.2. Geographical Distributions of June-August Mean Rainfall (mm d⁻¹) over the Amazon Basin

Actual Observations (a-d)

e) 180-km model   f) 120-km model   g) 60-km model   h) 20-km MRI-

Source: Figure generated for the report by Kitoh et al. 2009.
Note: Plots (a-d) correspond to data-sets of actual observations; (e) is 180-km model, (f) 120-km model, (g) 60-km model and (h) 20-km MRI-GCM model.

There are no distinct differences in large-scale patterns of precipitation with different horizontal resolutions. Figure 2.2 shows the June–August (JJA) mean precipitation climatology of the four observed SST datasets and the model at different scales. During this season under the current climate, a major rain area moves northward and large precipitation is found over northern South America while it is very dry over Northeastern Brazil and Southern Amazonia. Southern Brazil is covered with a rainy area extending from the South Atlantic. The Earth Simulator reproduces these rainfall distributions quite well.

Projection of Future Climate over the Amazon Basin

Rainfall

Projected changes at the end of the 21st century (2075–2099) were compared to the present (1979–2003). An overall pattern of precipitation change simulated by the 20-km and 60-km models is similar to that of ensemble means of CMIP3 models reported in IPCC (2007) with a large increase in the tropics, an increase in the mid- and high latitudes and a decrease in the subtropics. Four different boundary conditions (sea surface temperature change experiments) were undertaken. Figure 2.3 shows the annual mean precipitation change between the present and the future at a 60-km resolution for four different SSTs, that is, CMIP3 ensemble SST, CSIRO SST, MIROCH SST, and MRI SST. The general pattern of precipitation changes is similar and thus robust among different SSTs used,

Figure 2.3. Annual Mean Precipitation Changes (mm d⁻¹) between the Present and the End of the 21st Century for 60-km Resolution for Different Sea Surface Temperatures

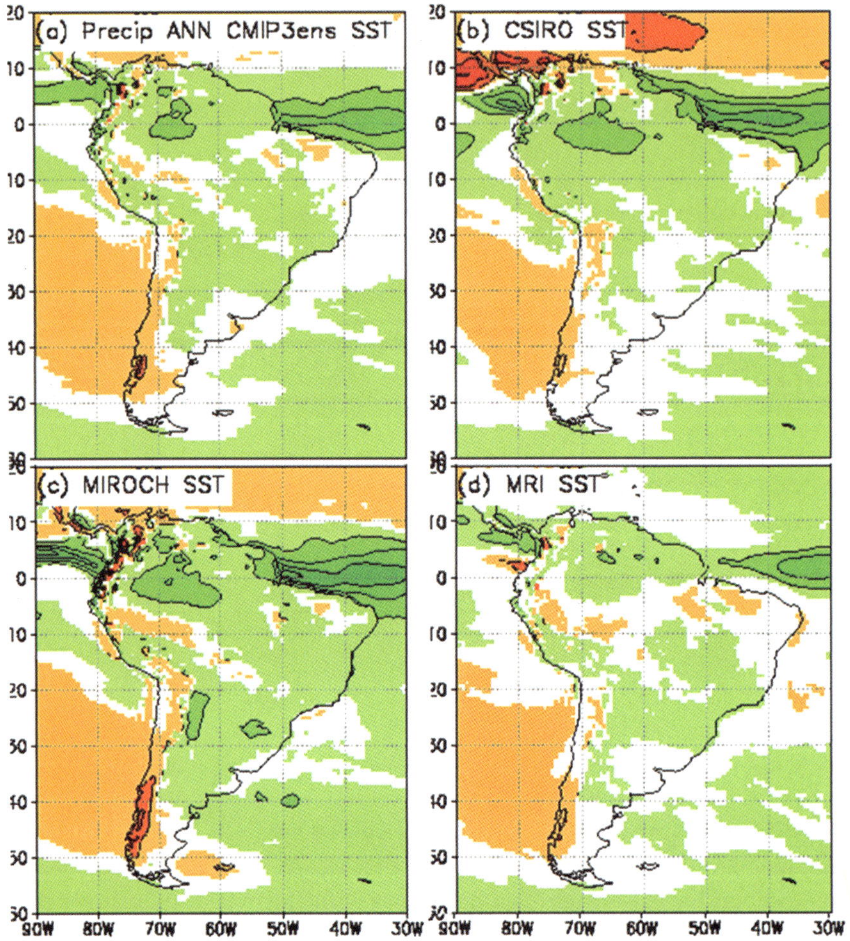

Source: Figure generated for the report by Kitoh et al. 2009.
Note: Figure presents data from (a) CMIP3 ensemble SST, (b) CSIRO SST, (c) MIROCH SST, and (d) MRI SST. Areas statistically significant at 95 percent level are colored. Contour interval is 1 mm d⁻¹.

with a large increase in precipitation over the ITCZ and the Northwestern Amazon and reductions in the Northeast and in the South.

Evaporation, Soil Moisture, and Surface Runoff

Due to the temperature increase, seasonal mean changes in evaporation, soil wetness at the uppermost layer, and surface runoff between the present and the future are anticipated. Evaporation increases are projected throughout the year over almost all of South America. An exception is Northeastern Brazil where evaporation is projected to decrease in the dry season (June, July, August and September, October, November). This is associated with drier soil over that region (figure 2.4). Drier soil is not restricted to Northeastern Brazil but is projected to occur over most of the continent, particularly in the dry season. Even in the wet season (DJF), the Amazon is expected to be much drier at the uppermost layer of the soil.

Significant changes in runoff are predicted for some regions. In September, October, November, and December, January, and February, runoff in the Western Amazon will decrease while it increases in Eastern and Southern Brazil. In March to May, Northwestern Amazonia experiences more runoff. Further analyses should be conducted to better ascertain regional hydrological changes.

Extreme Events

Global warming will result not only in changes in mean conditions but also in increases in the amplitude and frequency of extreme precipitation events. Changes in extremes are more important for the visualization of adaptation measures. Two extreme indexes for precipitation are used to illustrate changes in precipitation extremes over the Amazon, one for heavy precipitation and one for dryness. Figure 2.5 shows the changes in the maximum five-day precipitation total in mm (RX5D) for the 60-km and 20-km resolution. Throughout the basin, RX5D is projected to increase in the future. The largest RX5D increases (rainfall intensification) are found over the Northwestern Amazon and Southern Brazil. At a higher resolution (20-km), the model projects even larger increases in RX5D by the end of the century.

Likewise, figure 2.6 shows the changes in maximum number of consecutive dry days (CDD). A "dry day" is defined as a day with precipitation of less than 1 mm d^{-1}. Large CDD change is projected over the entire basin.

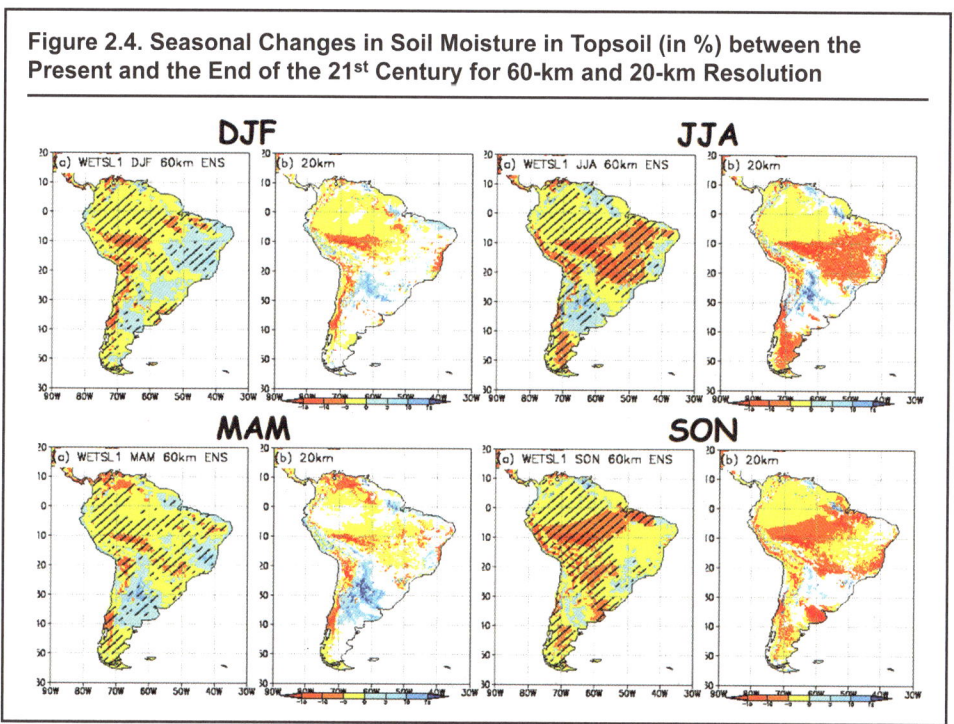

Figure 2.4. Seasonal Changes in Soil Moisture in Topsoil (in %) between the Present and the End of the 21st Century for 60-km and 20-km Resolution

Source: Figure generated for the report by Kitoh et al. 2009.
Note: For 60-km, areas statistically significant at 95 percent level are colored, and areas where all four different SST experiments show consistent changes in sign are hatched. For 20-km model, areas statistically significant at 90 percent level are colored. Contour interval is 1 mm d^{-1}.

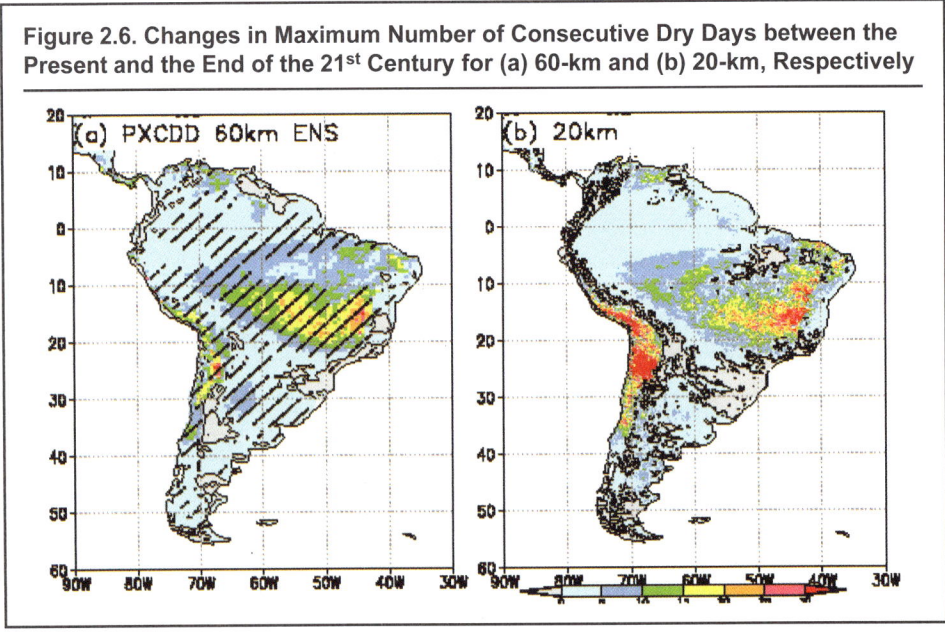

Figure 2.5. Changes in Maximum Five-Day Precipitation Total (mm) between the Present and the End of the 21st Century for (a) 60-km and (b) 20-km, Respectively

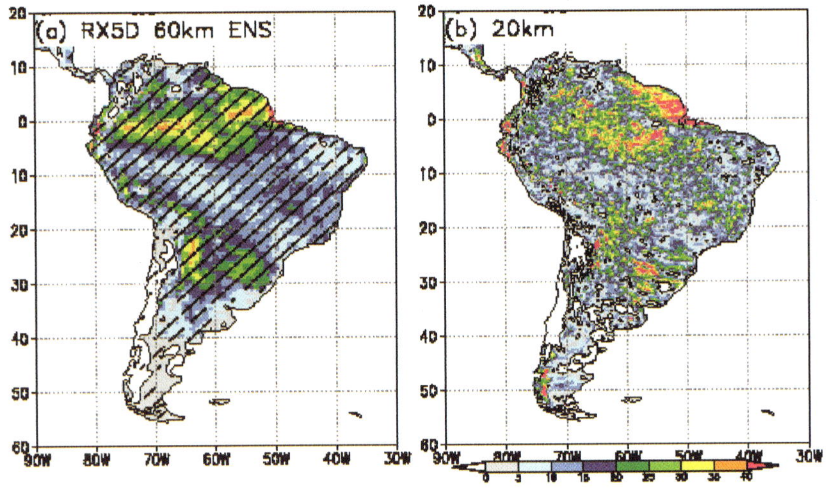

Source: Figure generated for the report by Kitoh et al. 2009.
Note: For 60-km model, areas where all four different SST experiments show consistent changes in sign are hatched. Zero lines are contoured.

In the present-day simulation, some intermittent rain occurs in the dry season, which is approximately from June to August over the Amazon. This intermittent rain in the dry season is confirmed by ground-station data. In the future climate simulation, this rainfall

Figure 2.6. Changes in Maximum Number of Consecutive Dry Days between the Present and the End of the 21st Century for (a) 60-km and (b) 20-km, Respectively

Source: Figure generated for the report by Kitoh et al. 2009.
Note: For 60-km model, areas where all four different SST experiments show consistent changes in sign are hatched. Zero lines are contoured.

during the dry season vanishes (figure 2.7), and thus stronger CDD signals are induced. Kamiguchi et al. (2006) analyzed future increases in CDD over the Amazon using a former experiment dataset with the 20-km mesh model. They found that in the present-day simulation, when rain takes place in the dry season, equatorial easterly low-level wind from the ITCZ hits the Andes and deviates to the south, bringing scattered clouds to the

Figure 2.7. Ten-Year Run of Daily Rain during the Dry Season at Armiquedes (present [blue] vs. future [red])

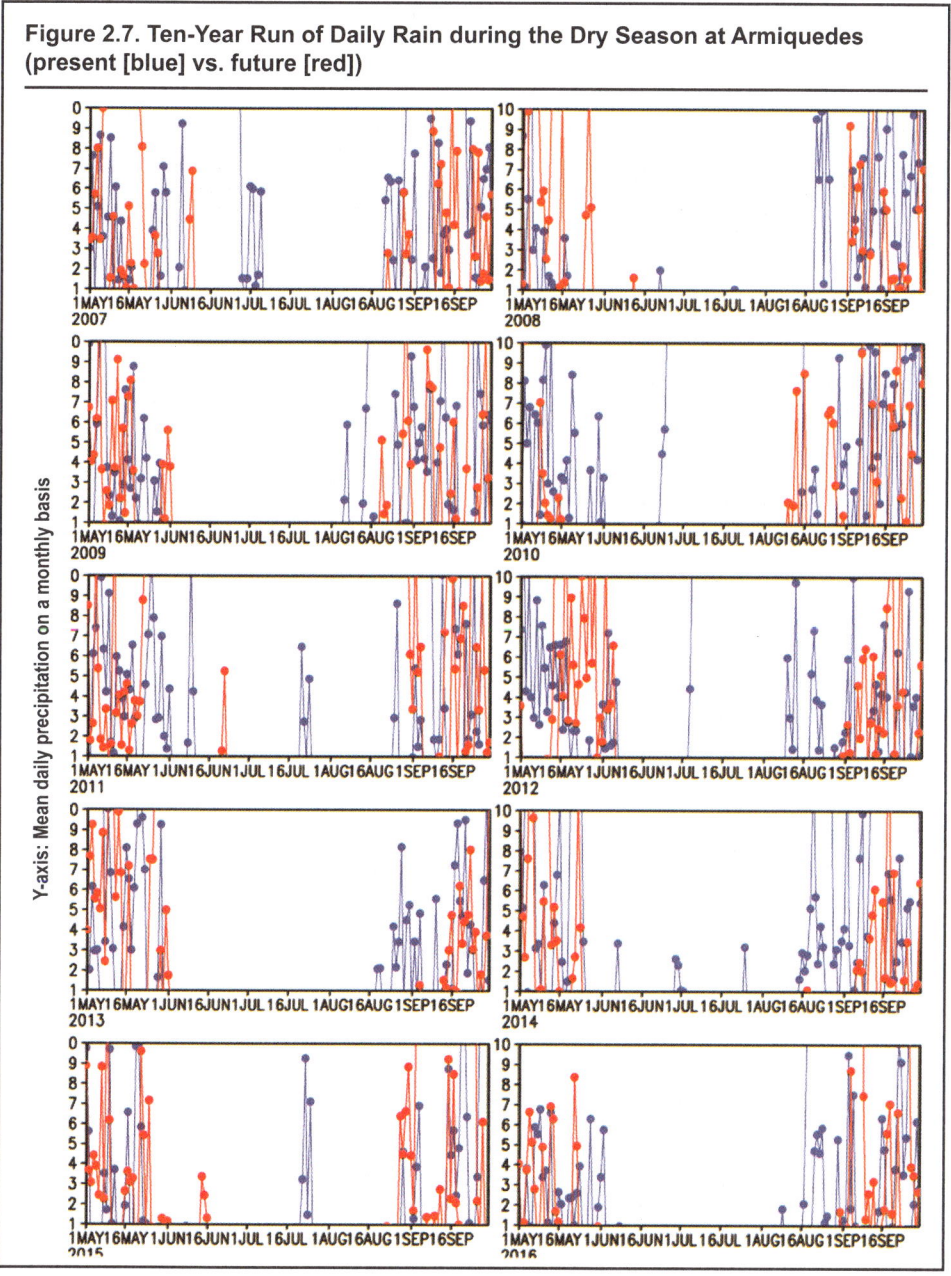

Source: Figure generated for the report by Kitoh et al. 2009.
Note: Present includes current decade; future, a decade at the end of the century. The graphs indicate that intermittent dry-season rainfall in the Amazon basin is projected to disappear.

middle of the Amazon. In the future climate, a weak Walker circulation[2] associated with the El Niño-like SST changes and a weak Atlantic anticyclone contribute to the weakening of the equatorial easterly wind and the suppression of rainfall over Amazon.

The analysis of future increases in CDD over the Amazon is probably one of the most important findings of the application of the Earth Simulator to the future climate in the basin and could have significant implications for the forests' resilience (increased vulnerability) to drought periods.

Impact on River Stream Flow

Using the runoff data derived from rainfall projections of the Earth Simulator, the stream flow of large rivers was calculated. The analysis used a "GRive T" river model.[3] In the present-day simulation, large rivers such as the Amazon and Parana are well represented by this model. The seasonal changes in the Amazon basin's discharges have been analyzed. The study selected the Obidos observation site situated between Northwestern Amazonia and Eastern Amazonia. The model simulates well the present seasonal cycle of stream flow. The end-of-century projection shows that the annual range of stream flow becomes larger (higher peaks and lower nodes), implying more floods in the wet season and more pronounced droughts in the dry season. This is consistent with the findings for RX5D and CDD.

Finally, the analysis included the impacts on seasonal flows of selected rivers in the Amazon. Figure 2.8 (left) shows the seasonal change in the discharges of the Rio Madeira by the 20-km mesh model at the nearest grid point to the observation site at Porto Velho (64.5°W, 9.3°S). The dashed line is for the present and the solid line is for the end of the 21st century. Long-term mean values (indicating the range of interannual standard deviations by shadings) of river discharge at observation points (Porto Velho) by the Global River Discharge Center (GRDC) are shown together. This point is situated at the exit of the river from Southern Amazonia. The 20-km mesh model reasonably

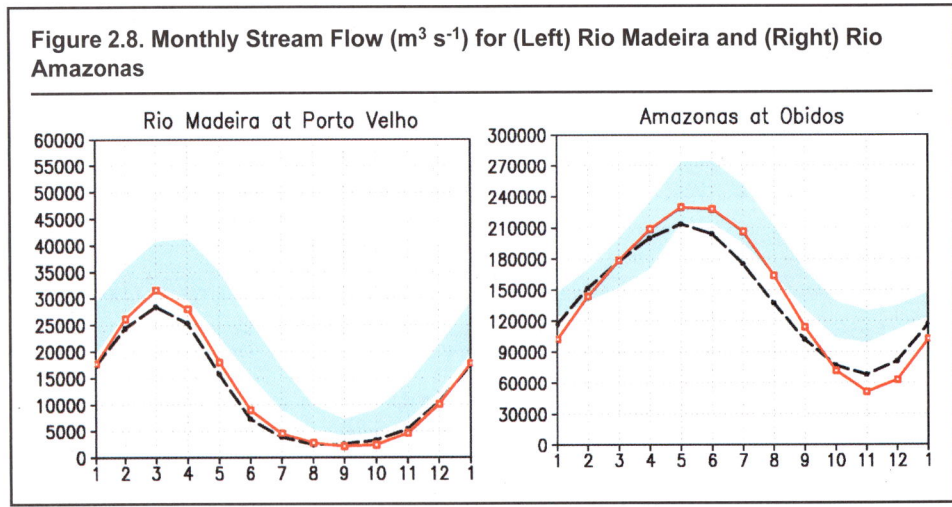

Figure 2.8. Monthly Stream Flow (m³ s⁻¹) for (Left) Rio Madeira and (Right) Rio Amazonas

Source: Figure generated for the report by Kitoh et al. 2009.
Note: Blue shadings denote the observed measurements by GRDC with interannual standard deviations. Observed data are at Porto Velho (64.5°W, 9.3°S) and at Obidos (55.8°W, 2.5°S). The heavy dashed line and the heavy red solid line denote the simulated stream flow for the present (1979–2003) and in the future (2075–2099), respectively.

reproduces the present seasonal cycle of stream flow at this point with maximum flow in March and minimum flow during August–October. The model simulation does not include any anthropogenic effects in river flows (diversions, reservoirs). It is projected that in the future, compared to today, stream flow will increase at the peak flow season (February–April) while it decreases during the late dry season.

Figure 2.8 (right) shows the seasonal change in the discharges of the Amazon River by the 20-km mesh model at the nearest grid point to the observation site at Obidos (55.8°W, 2.5°S). This point is situated between Northwestern Amazonia and Eastern Amazonia. Again, the model well reproduces the present seasonal cycle of stream flow at this point, but with some underestimation during low flow season.

It is projected that future stream flows in the Amazon basin will increase in the high-flow season and decrease in the low-flow season. It is also projected that the peak month of the high-flow season may be delayed and the high-flow season may become longer. It is noted that the annual range of stream flow becomes larger, implying more floods in the wet season and droughts in the dry season. For example, as a result of the intensification of the precipitation cycle and the increase in evaporation, the Amazon River projects higher amplitude in flows from 200,000 to nearly 230,000 m^3/s for the peak flow and from 80,000 to 60,000 m^3/s for the lower flows.

Notes

1. Today's emissions trajectory is already well above the A1B scenario. Therefore, this scenario may no longer represent a plausible future.
2. The Walker circulation is an ocean-based system of air circulation that influences weather on the Earth. The Walker circulation is the result of a difference in surface pressure and temperature over the western and eastern tropical Pacific Ocean (UCAR 2008. Online at: http://www.windows.ucar.edu/tour/link=/earth/Atmosphere/walker_circulation.html).
3. GRiveT: Global Discharge model using Total Runoff Integrating Pathways (TRIP), the 0.5 x 0.5 version with global data for discharge channels; Nohara et al. (2006). The river runoff assessed in the land surface model is horizontally interpolated as external input data into the TRIP grid so that the flow volume is saved. A similar analysis made for the Magdalena river in Colombia has recently been published (Nakaegawa and Vergara, 2010)

CHAPTER 3

Assessment of Future Rainfall over the Amazon Basin

The modeling at very high resolution by the Earth Simulator, although remarkable in its level of detail and unique in its ability to discern extreme events, represents the outcome of only one model applied to one scenario. It does confirm the expected intensification of the water cycle and, through the use of the 60 km ensembles, provides a level of robustness to its results. However, there remains a need to reduce uncertainties on modeling results over Amazonia, in particular as it concerns the projections related to rainfall, where there is considerable variance in modeling results.

Some GCMs predict increasing rainfall over the basin, while others predict drying (Li et al. 2006). In at least one major model the rainfall reduction is so severe as to undermine the status of the rainforest—leading to "dieback" (Cox et al. 2000; Cox et al. 2004; Cox et al. 2008). This chapter describes the development of *Probability Density Functions (PDFs)*[1] for future Amazonian rainfall, based on the projections produced by the 24 GCMs in use by the IPCC (and included in the CMIP3 archive) as a possible mechanism to reduce uncertainties in projected rainfalls in the region (table 3.1). These projected changes in rainfall represent the anticipated impact from the global emission of GHGs.

Table 3.1. GCMs in the CMIP3 Archive

Model Identifier	Model Name	Model Identifier	Model Name
a	bccr_bcm2.0	m	ingv_echam4
b	ccma_cgcm3_1	n	inmcm_3_0
c	ccma_cgcm3_1_t63	o	ipsl_cm4
d	cnrm_cm3	p	miroc3_2_hires
e	csiro_mk3.0	q	miroc3_2_medres
f	csiro_mk3.5	r	miub_echo_g
g	gfdl_cm2.0	s	mpi_echam5
h	gfdl_cm2.1	t	mri_cgcm2_3_2A
i	giss_aom	u	ncar_ccsm3_0
j	giss_model_e_h	v	ncar_pcm1
k	giss_model_e_r	w	ukmo_hadcm3
l	iap_fgoals1_0_g	x	ukmo_hadgem1

Source: Table generated for the report by Cox and Jupp 2009.

These GCMs produce very different predictions of rainfall in the selected geographical domains and also very different simulations of current climate in these regions (see figure 3.1). Most GCMs tend to overestimate rainfall in Northeastern Brazil but underestimate rainfall in the other four regions. The trends in 21st century rainfall vary from about +1 mm/day/century (e.g., Model "o" in the EA and NEB regions) to a drying of -2mm/day/century (e.g., Model "w" in the EA).

Figure 3.1. Simulation of Mean Annual Rainfall in the 20th Century (x-axis) and Rainfall Trend in the 21st Century (y-axis) in the Five Study Regions of Amazonia

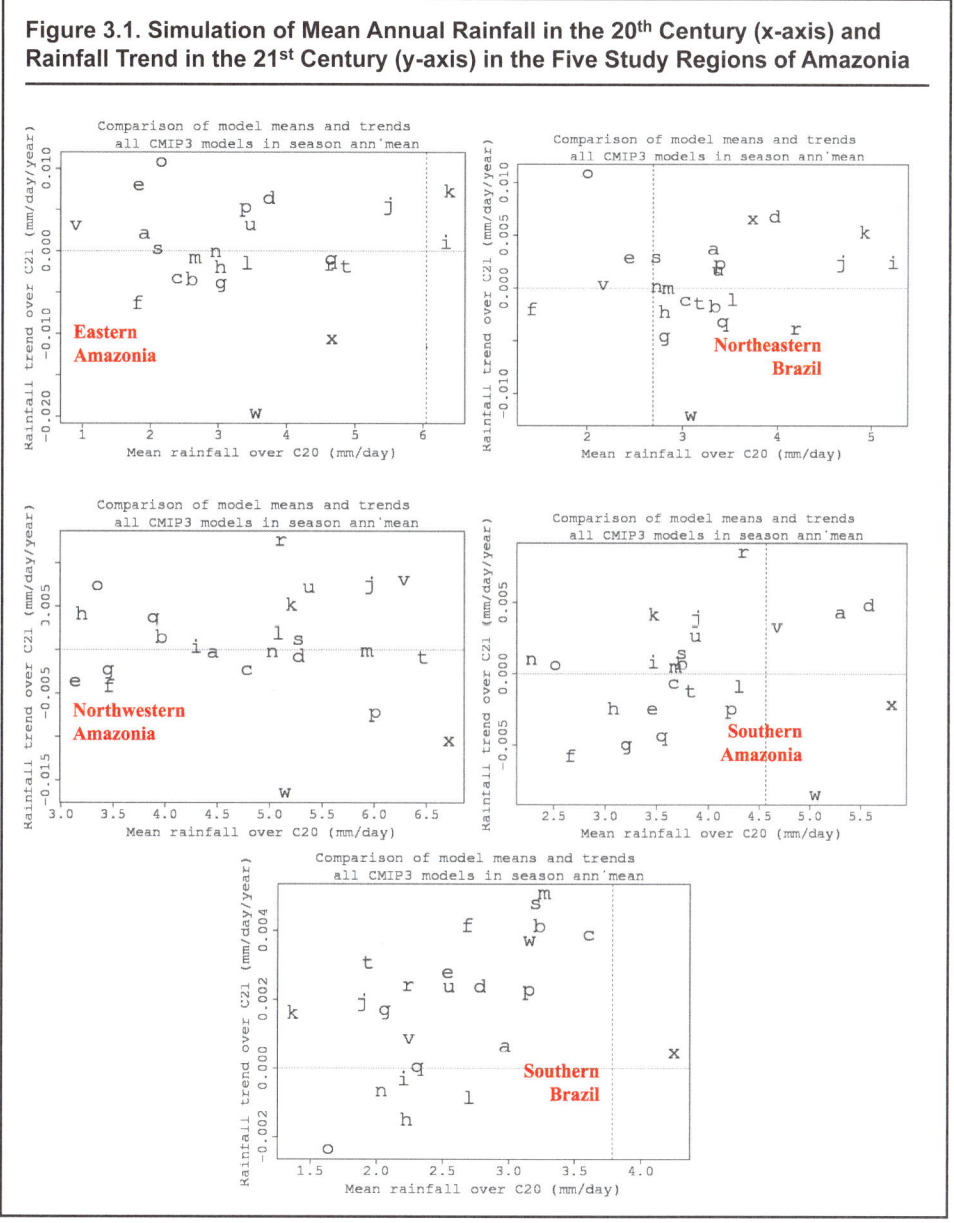

Source: Figure generated for the report by Cox and Jupp 2009.
Note: The vertical dotted line is the observed rainfall in the 20th century. The CMIP3 GCMs are labeled as in the table above.

Method for Estimating Probability Density Functions

One way to address uncertainty in future rainfall is to weight the various model projections based on the ability of each model to produce key aspects of the observed climate. In this way more robust predictions may be found by emphasizing the results from more realistic models and de-emphasizing the results produced by less realistic models.

The approach followed is to construct a probabilistic prediction based on a *weighted* sum of the predictions of individual GCMs, using a Bayesian approach.[2] The weight assigned to each GCM will be referred to as *the probability of the model* and will generate a probability density function (PDF) over the set of models.[3]

This procedure can be used to estimate PDFs for future rainfall in each of the five regions, using simulations produced by the 24 GCMs available in the archive of the *Coupled Model Intercomparison Archive Project (CMIP3)*. Two approaches are used to weight the respective model outputs, the first based purely on the rainfall simulated by each GCM (see below), and the second using the observed correlation between Amazonian rainfall and sea surface temperature anomalies in the Atlantic and Pacific Oceans (see below).

Models Weighted According to Rainfall Projections

There is a need to capture errors in the simulation of the mean climate and its variability, so that the models are penalized for both the bias in the mean rainfall simulation and the Kolmogorov-Smirnov statistic[4] of the bias-corrected rainfall. Using these metrics and the Bayesian approach described in the notes at the end of the chapter, relative model weightings for each season ("DJF," "MAM," "JJA," "SON") are derived, as well as for the entire calendar year ("ann_mean"), for each geographical domain. The annual mean weights, which have subsequently been used to weight the projections of the LPJ dynamic global vegetation model (see chapter 4), are shown in figure 3.2.

The approach utilized results in the selection of some models ahead of others, with just a minority of models being favored in each region. The number of models with more than the mean weighting is between six and nine depending on the region, with most of the other models receiving very low weights. However, the most trusted models vary significantly between the regions, and no single model has above-average weighting for all five regions. In any one region, it is also unusual for a given model to simulate rainfall accurately in all four seasons.

However, the procedure distinguishes strongly among the models. Figure 3.3 shows example PDFs and Cumulative Distribution Functions (CDFs) for Southern Amazonia, in each case for the 2001–2031 (black) and 2068–2098 (red) periods. The figure also shows the difference between the prior distributions (dotted lines), which were derived by assuming all models to be equally likely, and the posterior distributions (continuous lines), which were derived using the outlined procedure. In most cases weighting the models by their relative abilities to produce the current rainfall leads to sharper PDFs. In some cases the weighting also shifts the most likely future rainfall significantly.

The regions and seasons where significant changes in the rainfall PDF are predicted to occur during the 21st century include: Northwestern Amazonia, which is predicted to become wetter in the DJF and MAM periods; and Southern Amazonia, which has an increased probability of 2005-like drought conditions in the SON pre-wet season (see bottom-left panel of figure 3.3).

Figure 3.2. Relative CMIP3 Model Weightings Based On Simulation of Annual Rainfall (Mean and Variability) for the Five Land Regions of Amazonia

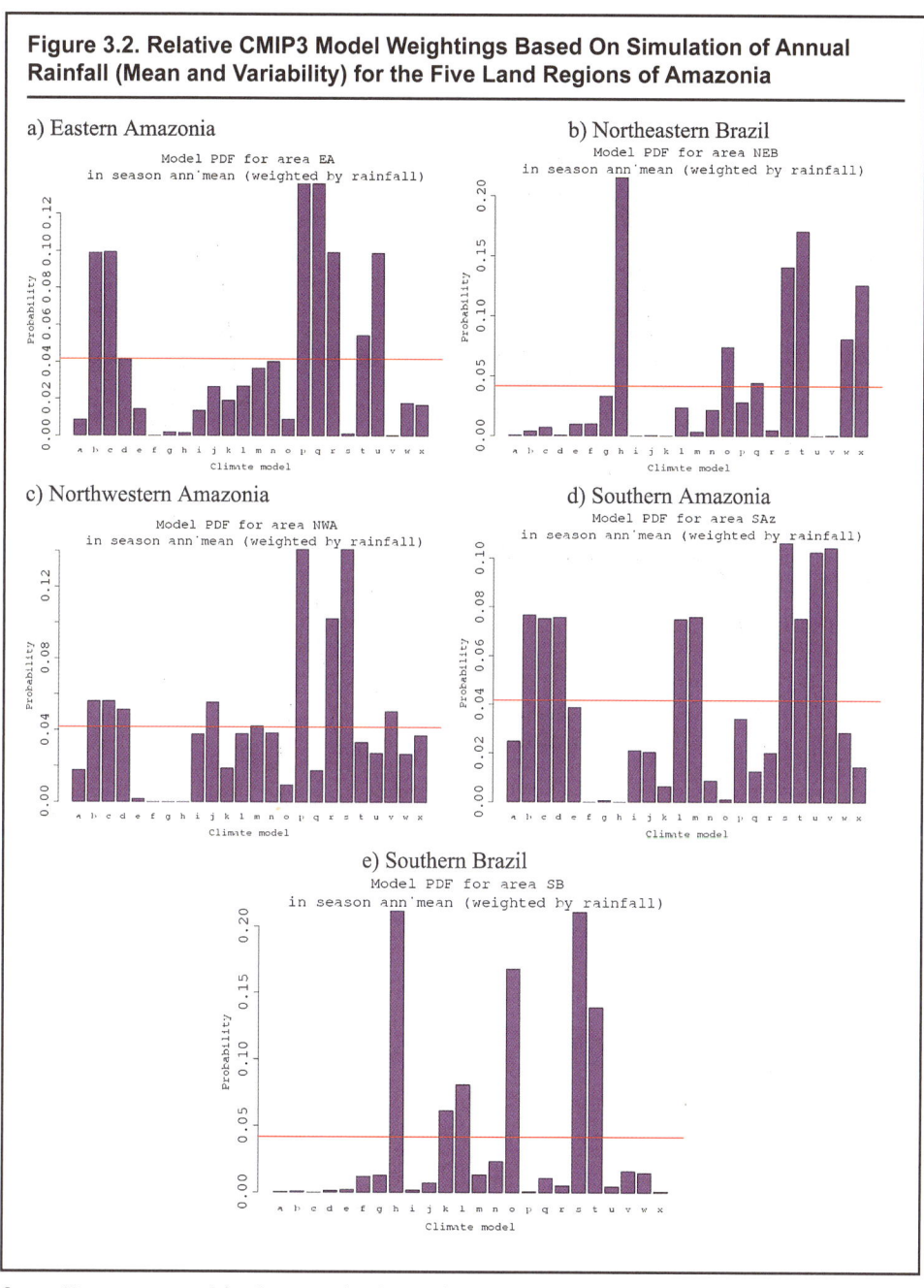

a) Eastern Amazonia

b) Northeastern Brazil

c) Northwestern Amazonia

d) Southern Amazonia

e) Southern Brazil

Source: Figure generated for the report by Cox and Jupp 2009.
Note: The horizontal line shows the expected probability if all models are equally likely. The CMIP3 GCMs are labeled as in table 1.1.

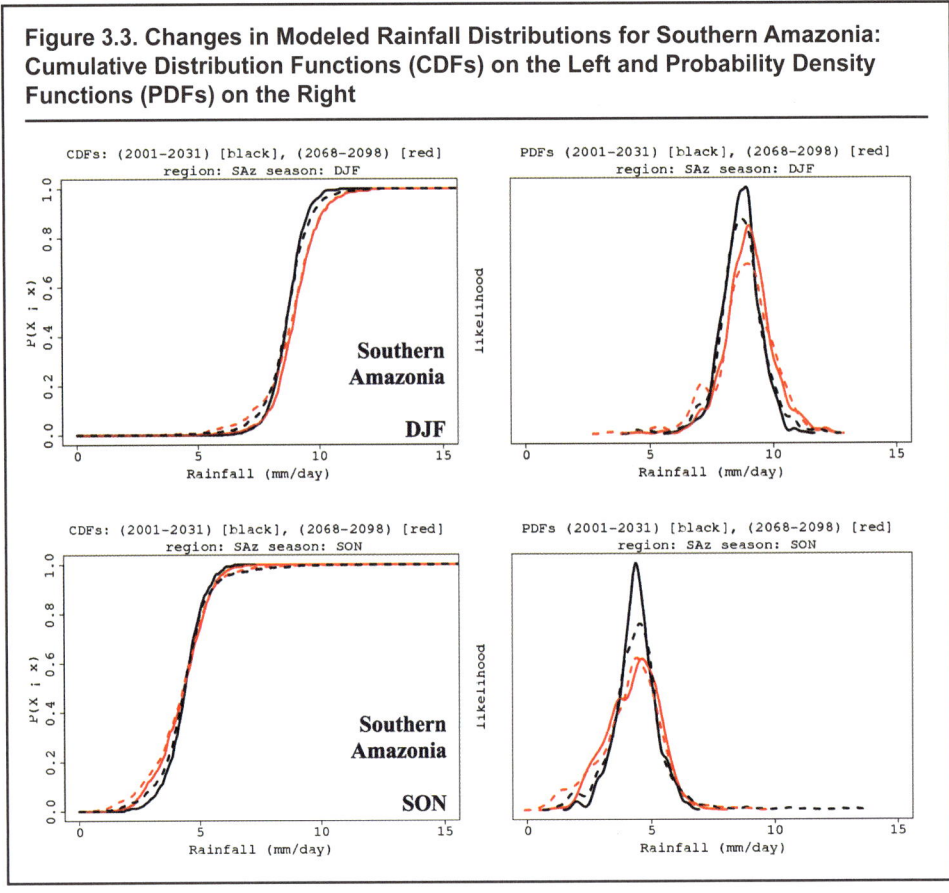

Figure 3.3. Changes in Modeled Rainfall Distributions for Southern Amazonia: Cumulative Distribution Functions (CDFs) on the Left and Probability Density Functions (PDFs) on the Right

Source: Figure generated for the report by Cox and Jupp 2009.
Note: Black lines represent the 2001–2030 period and red lines show 2068–2098. The dotted lines are based on a uniform prior in which all model projections are considered equally likely, and the continuous lines are from the Bayesian weighting procedure based on the rainfall simulations of each model. Note the increased probability of extreme wet conditions in DJF, and extreme (2005-like) dry conditions in SON.

The probability of annual rainfall being less than 3 mm/day was also calculated for each of the five regions. The results are summarized in table 3.2.

Table 3.2. Probability of Annual Rainfall Being Less than 3mm per Day for Each of the Five Study Regions of Amazonia

Region	Probability 2001–2030	Probability 2068–2098
EA	0%	0.7%
NEB	80%	76%
NWA	0%	0%
SAz	0%	0.1%
SB	1.1%	6.8%

Source: Table generated for the report.

Models Weighted According to Sea Surface Temperature Indexes

Rainfall in Amazonia is also known to be sensitive to seasonal, interannual, and decadal variations in sea surface temperatures (Marengo 2004). For example:

- Warming of the tropical East Pacific during El Niño events suppresses wet-season rainfall through modification of the (East-West) Walker Circulation and through the Northern Hemisphere extra-tropics (Nobre and Shukla 1996). El Niño-like climate change (Meehl and Washington 1996) has similarly been shown to influence annual mean rainfall over South America in GCM climate change projections (Cox et al. 2004; Li et al. 2006);
- Warming of the tropical North Atlantic relative to the South leads to a north-westward shift in the Intertropical Convergence Zone (ITCZ) and compensating atmospheric descent over Amazonia (Liebmann and Marengo 2001; Fu et al. 2001; Cox et al. 2008). For Northeastern Brazil the relationship between the North-South Atlantic SST gradient and rainfall is sufficiently strong to form the basis for a seasonal forecasting system (Folland et al. 2001). Variations in SSTs in the tropical Atlantic and Pacific contribute in different ways to rainfall variability in the regions of Amazonia (Cox and Jupp 2008).

Multiple linear regressions of rainfall are reported in the various land regions of South America against indexes of the Pacific East-West SST gradient (PEWG) and the Atlantic North-South SST gradient (ANSG), based on observational data (New et al. 2000; Rayner et al. 2003).[5] Rainfall is correlated in the five selected regions against the *ANSG*, which is the SST difference between the tropical North Atlantic [75°W–30°W, 15°N–35°N] and the tropical South Atlantic [40°W–15°E, 25°S–5°S]; and the *PEWG* which is the SST difference between the tropical East Pacific [150°W–90°W, 5°S–5°N] and the tropical West Pacific [120°E–180°E, 5°S–5°N]. Linear correlations were carried out for each season (December to February=DJF; March to May=MAM; June to August=JJA; September to November=SON) and also for the annual mean.

The computed correlation coefficients and their uncertainties (given as a standard deviation) were calculated and are available upon request from the authors. From these, some significant impacts of *ANSG* and *PEWG* on rainfall in the designated regions were inferred. For example:

- An anomalously positive ANSG (i.e., 2005-like conditions in which the North Atlantic warms relative to the South) leads to reduced rainfall year-round in Northeastern Brazil, and from July to November in Eastern Amazonia.
- An anomalously positive PEWG (i.e., El Niño-like conditions in which the East Pacific warms relative to the West) leads to reduced rainfall in the East (Northeastern Brazil and Eastern Amazonia) but to increased rainfall in the South (especially in Southern Brazil).

General Circulation Model Simulation of Current and Future Sea Surface Temperature Indexes

Given the observed correlation between the ANSG and PEWG indexes and Amazonian rainfall, the simulation of these indexes by the models is likely to have relevance to the simulation of rainfall in the future. The left panel of figure 3.4 shows the annual mean

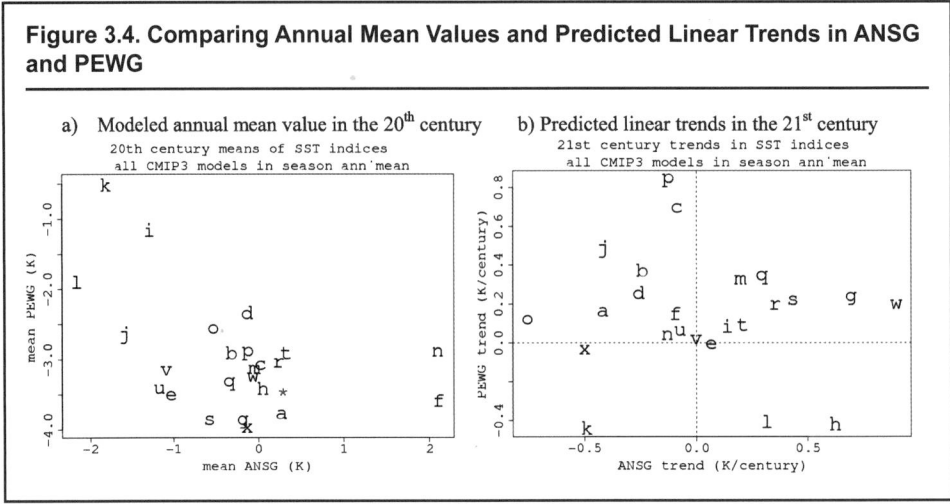

Figure 3.4. Comparing Annual Mean Values and Predicted Linear Trends in ANSG and PEWG

a) Modeled annual mean value in the 20th century   b) Predicted linear trends in the 21st century

Source: Figure generated for the report by Cox and Jupp 2009.
Note: Model labels can be found from the table 3.1. In the panel a) observed values shown by red asterisk.

ANSG and PEWG values as simulated by each of the CMIP3 GCMs, and compares them to the values estimated from observations (red asterisk). Few models simulate both of these indexes well, with most models producing a negative bias in ANSG (South Atlantic too cold relative to North), and a positive bias in the PEWG (East Pacific too warm relative to the West).

The right panel of figure 3.4 shows the 21st century trends in PEWG and ANSG from each model. Most models seem to produce a warming of the East Pacific relative to the West (El Niño-like pattern), but there is no agreement about the sign of the trend in the ANSG, with some models presuming strong warming of the North Atlantic relative to the South (e.g., Model "w") and others producing the opposite (e.g., Model "o").

The observed data indicate that more positive ANSG is associated with drying in all regions aside from Southern Brazil. This observed difference between the different GCMs is very likely to impact the simulated trends in rainfall.

Probability Density Functions for Future Sea Surface Temperature Indexes

In this subsection weighting factors for each model are estimated based on simulation of the annual mean ANSG and PEWG in the 20th century.[6] Figure 3.5 shows the weighting factors derived through comparison to ANSG data alone (left panel) and PEWG data alone (right panel). Six of the 24 models achieve above-average weighting on the ANSG (with model "w" being the highest weighted), and nine achieve above-average weighting for PEWG but only one significantly so (model "o").

Based on these weightings, estimates can be made of CDFs and PDFs for the ANSG and PEWG; these are shown in figures 3.6 and 3.7. The weighting produces a climate change signal with a suggestion that the most likely ANSG values will become more negative through the 21st century, which would make four of the five study areas wetter (see table 3.1). However, it also increases the probability of positive (2005-like) anomalies in the ANSG.

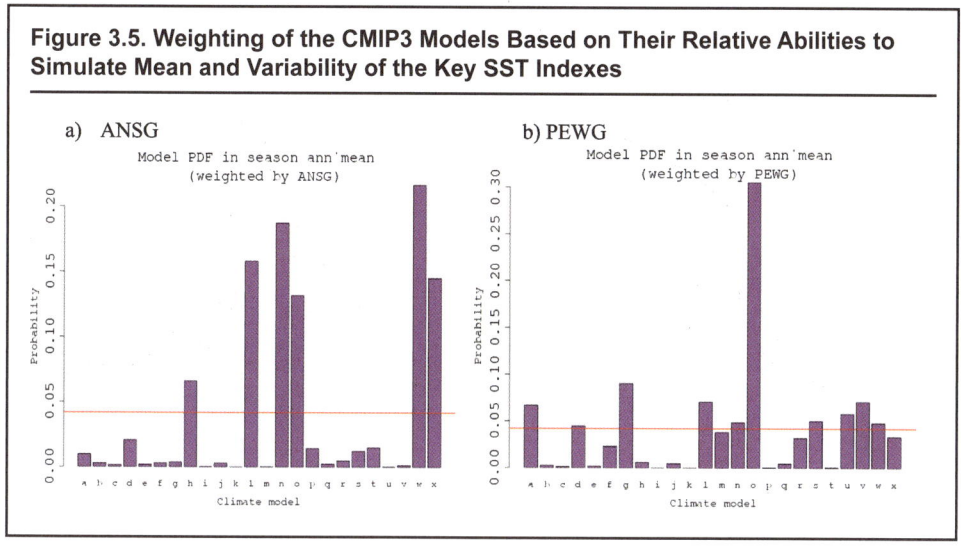

Figure 3.5. Weighting of the CMIP3 Models Based on Their Relative Abilities to Simulate Mean and Variability of the Key SST Indexes

Source: Figure generated for the report by Cox and Jupp 2009.

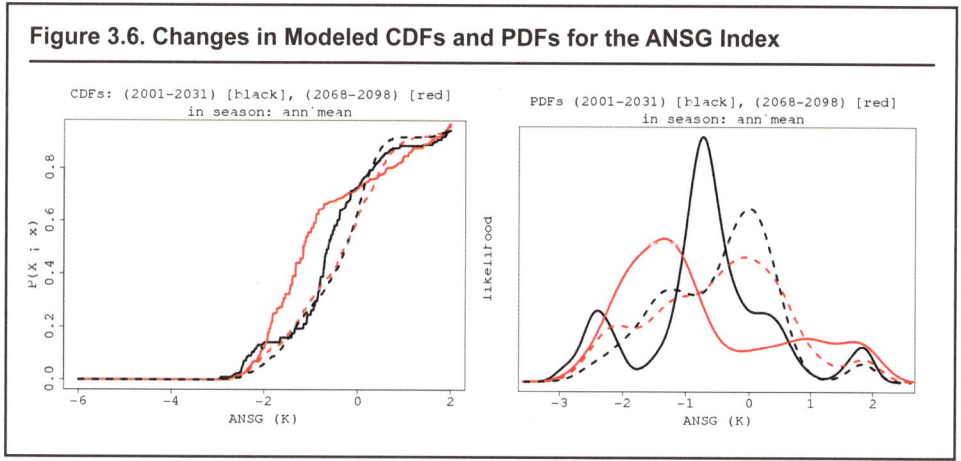

Figure 3.6. Changes in Modeled CDFs and PDFs for the ANSG Index

Source: Figure generated for the report by Cox and Jupp 2009.
Note: Black lines represent the 2001–2030 period and red lines show 2068–2098. The dotted lines are based on a uniform prior in which all model projections are considered equally likely, and the continuous lines are from the Bayesian weighting procedure based on the ANSG simulations of each model.

There is a suggestion of slight increases in PEWG through the 21st century (figure 3.7), which are consistent with the El Niño-like raw model trends shown in figure 3.4.

In summary, the PDF-based method calculated using the two procedures described in this chapter, although useful in detecting significant potential changes in rainfall in some regions, for some seasons, is not enough to support a blanket statement on whether the existing set of models predict a wet Amazon or a dry Amazon as a result of climate change. Some of the highest-ranked models indicate a tendency toward a wetter Amazon, particularly in the Northwest, but simultaneously point to an increased possibility of 2005-like events in Southern Amazonia. As for other regions, it is still difficult to discern a trend.

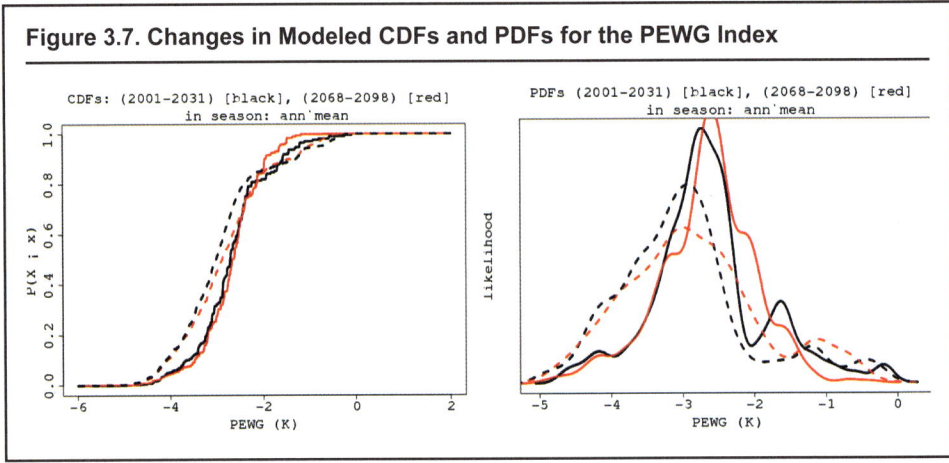

Figure 3.7. Changes in Modeled CDFs and PDFs for the PEWG Index

Source: Figure generated for the report by Cox and Jupp 2009.
Note: Black lines represent the 2001–2030 period and red lines show 2068–2098. The dotted lines are based on a uniform prior in which all model projections are considered equally likely, and the continuous lines are from the Bayesian weighting procedure based on the PEWG simulations of each model.

The analysis also indicates a significant increase in the probability of drought conditions in Southern Brazil (up from 1.1 percent in 2001–2030 to 6.7 percent in 2068–2098) or a shift from one 2005-like drought per century to one every seventeen years. Still, the overall uncertainties in future Amazonian rainfall remain significant.

Notes

1. PDFs define the probability that a particular variable (in this case Amazonian rainfall) falls within a given range. They can therefore be used to estimate the probability that the rainfall is less than some critical value that could lead to forest dieback.
2. Bayes's theorem allows the model probabilities to be modified each time one considers the ability of the models to simulate some relevant aspect of current climate (such as rainfall in each season). This updating of the PDF is achieved by comparing time series of past observations with time series of model simulations for each variable.
3. This updating of the PDF is achieved by comparing time series of past observations with time series of model simulations for each variable. In this study models are weighted based on their ability to simulate both the mean state and the variability (i.e., the statistical distribution) of current climate. In other words, the aim is to down-weight those models which simulate a climate whose mean value is far from the observed mean, or a climate whose statistical distribution is a poor fit to the observed distribution, even when any bias in the mean value has been corrected. The procedure can be summarized as follows:

   i.   assign equal probability to all models—a *uniform prior* PDF; choose a climatic variable of interest;
   ii. update the model PDF based on the fit between model simulations and observations for this variable;
   iii. treat the current model PDF as a new prior, and repeat steps (i) and (iii) as required;
   iv. obtain a final posterior PDF for the models.

4. The Kolmogorov-Smirnov statistic can be easily calculated using standard statistical software packages; it is used to measure distributional adequacy in a sample.
5. The land regions are shown schematically in Figure I.1 and are labeled as *Eastern Amazonia (EA), Northeastern Brazil (NEB), Northwestern Amazonia (NWA), Southern Amazonia (SAz) and Southern Brazil (SB).*
6. Models are penalized for any bias in the long-term mean as well as for errors in the simulation of variability (through the Kolmogorov-Smirnov statistic).

CHAPTER 4

Analysis of Amazon Forest Response to Climate Change

Introduction

The analysis in chapter 3 does not lead to a blanket statement regarding an increase or reduction in rainfall over the entire Amazon basin. However, it does indicate a tendency toward a wetter Amazon, in particular in the Northwest, but simultaneously indicates an increased possibility of dry events in Southern Amazonia and even more in Southern Brazil. Remarkably, these are the same trends identified through the use of the high-resolution Earth Simulation of the Amazon basin.

Beyond rainfall, however, there are other climate-induced impacts that need to be considered for their potential effect on both the structure and behavior of the forest. As described in chapter 1, these include the following:

- High probabilities for prolonged drought stress may lead to increasing physiological stress for trees, increased tree mortality, and thus carbon emissions.
- Increasing atmospheric CO_2 concentrations may alter the drought response of forests.
- Increasing temperatures may accelerate heterotrophic respiration rates and thus carbon emissions.
- Resulting changes in evapotranspiration and therefore convective precipitation could further accelerate drought conditions and destabilize the tropical ecosystem as a whole.

Strong impacts of drought on tropical forest biomass and structure are corroborated by field measurements and experiments. The observed responses of rainforests to drought events such as the 1997–1998 ENSO event range from high tree mortality of ~26 percent in a forest with seasonal rainfall (East Kalimantan; Van Nieuwstadt and Sheil 2005) to no mortality response in seasonally dry forests (Panama; Condit et al. 2004) and several intermediate responses (Williamson et al. 2000). During the Amazon drought in 2005, Phillips et al. (2009) measured strongly increased tree mortality and rather small declines in growth in the surviving trees. Mainly fast-growing light-wooded trees were affected by carbon starvation. Similar results have been found in the 1983 drought event in Panama, where mortality of different tree species increased (Condit et al. 1995). Large trees are especially sensitive to reduction in soil water below a critical threshold and react with increased mortality (Nepstad et al. 2007). Thus, in a drought event, trees may be killed selectively and thus species composition as well as canopy structure may be altered (Nepstad et al. 2007; Phillips et al. 2009).

Rainfall is one of the most important drivers of forest growth and survival, but the signal provided by different models has a large variance. The GCM model weightings described in chapter 3 provide information about the potential of the 24 climate models to reproduce the mean and the distribution of annual rainfall under current conditions. With the derived ranking of the climate models, there is the possibility to estimate the response of biomass and thus the risk of a potential forest dieback according to the climate models' probability.

In this chapter, the risk, mechanisms, and consequences of a possible dieback of undisturbed rainforests in Amazonia are assessed, using a state-of-the-art process-based vegetation model (LPJmL). The objectives were to estimate (a) the range of possible impacts of climate change on Amazonian forest ecosystems, and (b) the probability of a forest dieback, based on weighted probability density functions for the 24 climate models as estimated in chapter 3.

The Lund-Potsdam-Jena Managed Land Dynamic Global Vegetation and Water Balance Model

Ecosystem-level responses to changing environmental conditions (temperature, water availability, ambient CO_2) are the net outcome of multiple processes, such as photosynthesis, auto- and heterotrophic respiration, growth, competition, and mortality. The dynamics of ecosystem structure and vegetation composition are therefore highly nonlinear and depend strongly on geographical location and climate conditions. Observations of these processes and their net ecosystem impact exist from a range of sites throughout the tropical forest belt, and remote sensing data provide further regional-to-global integrative data. However, regionalizing these ecosystem responses for the main climatic and edaphic gradients, and extrapolating them under climate change and increasing CO_2 concentrations, require comprehensive numerical simulation models that include the main physiological, biogeochemical and stand-level processes, as well as vegetation dynamics due to competition and disturbance.[1]

For the present study, the LPJmL model has been applied to the Amazon region.[2] Since the goal of this task is the identification of forest responses to changing climate, all forests are assumed to be unaffected by land use change or deforestation. The LPJmL model is therefore used in its potentially natural vegetation mode. In the following chapter, the coupled climate-vegetation model CPTEC-PVM2.0 is then used to estimate the climate feedbacks from biome shifts and deforestation.

In LPJmL, vegetation processes are simulated for small areal units, the size of which is determined by available data. Typically, these units are cells in grids with mesh sizes of 0.5 degrees longitude and latitude.[3] The functional units of the model are plant functional types (PFTs) which can be conceived as plant species grouped by specific attributes controlling their physiology and dynamics. The projected PFTs in the Amazon basin are "tropical broadleaf evergreen," "tropical broadleaf raingreen," and "C_4 grass."

The vegetation in each grid cell is represented as a mixture of the three PFTs. Each of the PFTs covers a certain proportion of the modeled area, which is denoted as its "foliar projective cover" (FPC). Plant physiological and biogeochemical processes are simulated in a mechanistic way. Photosynthesis, water balance and maintenance respiration for each PFT are calculated on a daily time step. The assimilated carbon (NPP) is allocated to the different carbon compartments of the plant, such as leaves, wood, and roots, according to specified allometric constraints.

During this work, a new fire module (SPITFIRE, Thonicke et al. in review) has been incorporated into LPJmL; it simulates detailed climatic fire danger, ignition, spread, effects, and emissions of wildfires in natural vegetation caused by lightning (Thonicke et al. in review).

Simulation of Vegetation State in the Amazon Basin

To visualize the current state of the ecosystem and the level of forest degradation in the Amazon basin, three indicator variables directly derived from state variables of LPJmL are used:

- Forest cover (FC), which was calculated from foliar projective cover (FPC) of the woody PFTs and ranges between 0 and 100 percent;
- Biomass density (BD), which is the accumulated aboveground vegetation carbon in kg C m^{-2}; and,
- a vegetation classification scheme that describes forest types by their proportion of forest cover and biomass density of the natural stand within the grid cell (figure 4.1). In a simplified way,

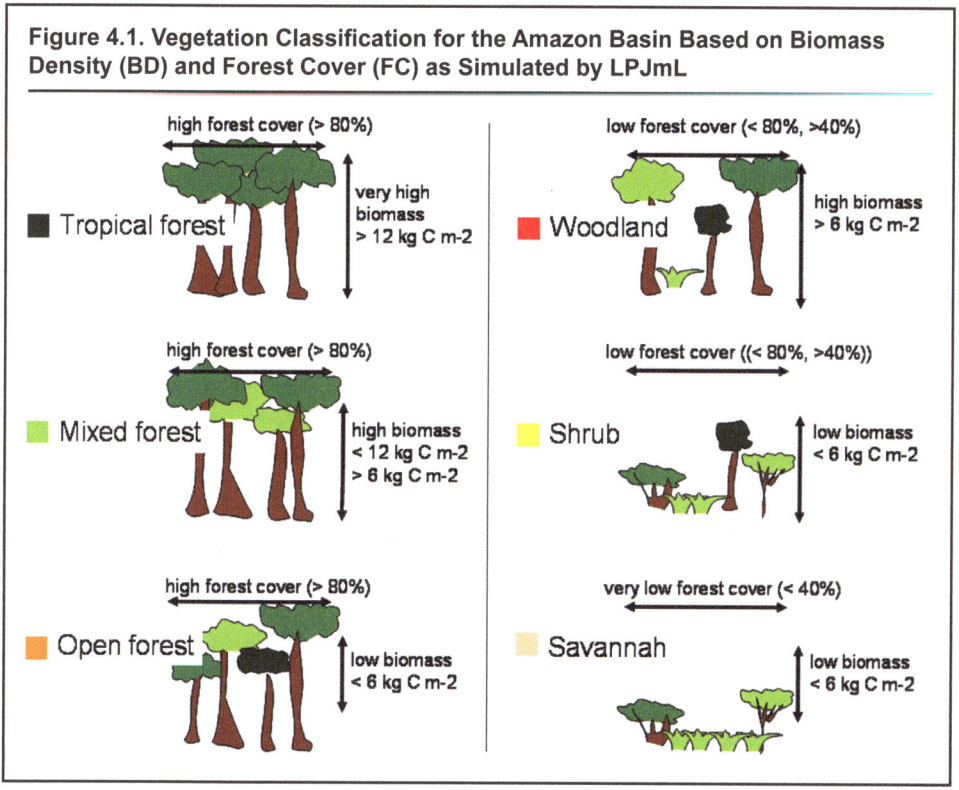

Figure 4.1. Vegetation Classification for the Amazon Basin Based on Biomass Density (BD) and Forest Cover (FC) as Simulated by LPJmL

Source: Figure generated for the report by Rammig et al. 2009.

- "tropical" (FC > 80, BD > 12) and "mixed" (FC > 80, BD < 12, BD > 6) forests are seen as "pristine" forest types, while
- "open forests" (FC > 80, BD < 6),
- "woodland" (FC <80, FC > 40, BD > 6),
- "shrubland" (FC < 80, FC > 40, BD < 6), and
- "savanna" (FC < 40, BD < 6) are either occurring naturally or can also be "secondary" degraded types, i.e., caused by climate or land use change. The threshold values for FC and BD were estimated following Alencar et al. (2006).

Under current climate conditions (1981–2000), the potential natural vegetation in the Amazon basin is dominated by tropical and mixed (deciduous) rainforests. LPJmL simulations of current vegetation are in accordance with other biome classification and potential vegetation maps (IBGE 1988; Salazar et al. 2007). The total estimated biomass (above- and belowground) ranges from 47 to 86 Pg C in the Amazon basin; this lies within the (broad) range of other simulation results and inventories which suggest values between 39 and 123 Pg C (Houghton et al. 2001; Cowling and Shin 2006; Malhi et al. 2006; Soares-Filho et al. 2006; Saatchi et al. 2007).

Response of Biomass to Projected Changes in Rainfall in the Different Geographical Domains

The biomass response as calculated from LPJmL based on the CMIP3 multi-model climatology has been ranked according to the PDFs for rainfall (calculated in chapter 3). Figures 4.2 to 4.6 show the estimated model weightings and the projected changes in biomass for the time period 2070–2100 versus 1970–2000 for the LPJmL-S1 (shallow roots and the SRES-A1B emission trajectory) simulations, while deliberately excluding the potential role of CO_2 fertilization.

Current knowledge and available data seem to indicate that, provided there are no limiting water and nutrient constraints, CO_2 fertilization plays a role in the growth of stands in temperate forests (this is particularly important for young forest stands). This assumption has been at the basis of current dynamic vegetation modeling.

However, under pronounced nutrient constraints, typical of poor soil conditions in the Amazon basin, there is substantial uncertainty that CO_2 fertilization may play such an effective role in tropical, mature forest ecosystems. Thus, in the absence of solid information (such as from ecosystem CO_2 fertilization experiments), the assumption that CO_2 fertilization will be significant in the Amazon cannot presently be used as a basis for sound policy advice.

The bar plot in the top panel of figures 4.2 to 4.6 shows the ranking of the 24 climate models for the respective geographical domain as calculated for the combined distribution over all seasons without the CO_2 effect. Models with the highest probability are best reproducing mean and distribution of rainfall patterns in the respective region. The box-and-whiskers plots in the panel below show the corresponding change in biomass, simulated by LPJmL under the SRES-A1B emission trajectory. The graphs show the difference between the average over the periods 2070–2100 and 1970–2000. Negative values indicate decrease and positive values indicate increase in vegetation cover and forest cover.

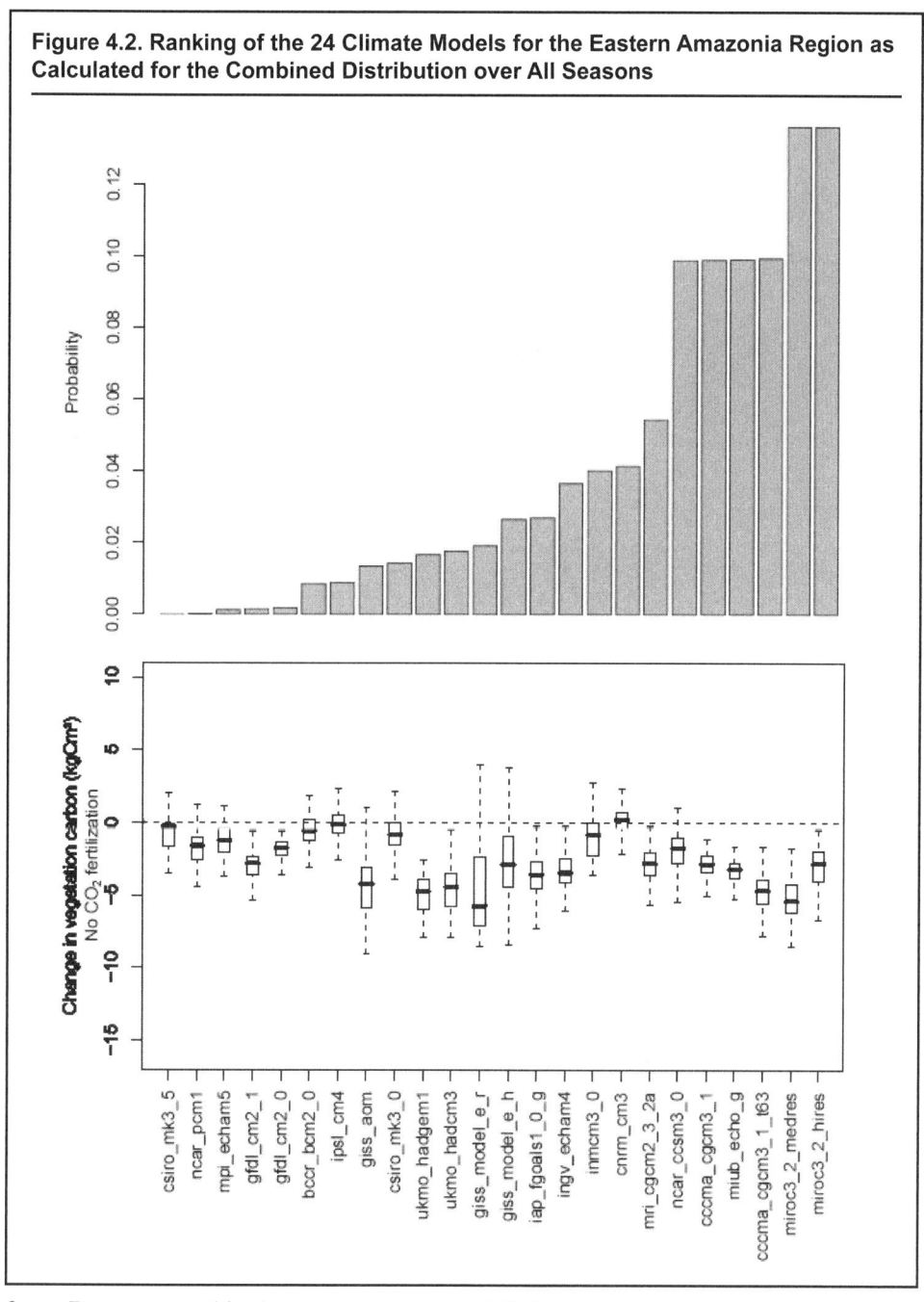

Figure 4.2. Ranking of the 24 Climate Models for the Eastern Amazonia Region as Calculated for the Combined Distribution over All Seasons

Source: Figure generated for the report by Rammig et al. 2009.
Note: Top panel, models with highest probability are best reproducing mean and distribution of rainfall patterns in this region.

For **Eastern Amazonia (EA),** two models rank best (miroc3_2_hires, miroc3_2_medres). Their climatologies lead to two slightly different responses of vegetation carbon projections under future conditions. Without CO_2 fertilization both models predict a decrease of between 2 and 5 kg C m^{-2}.

Figure 4.3. Ranking of the 24 Climate Models for Northwestern Amazonia as Calculated for the Combined Distribution over All Seasons

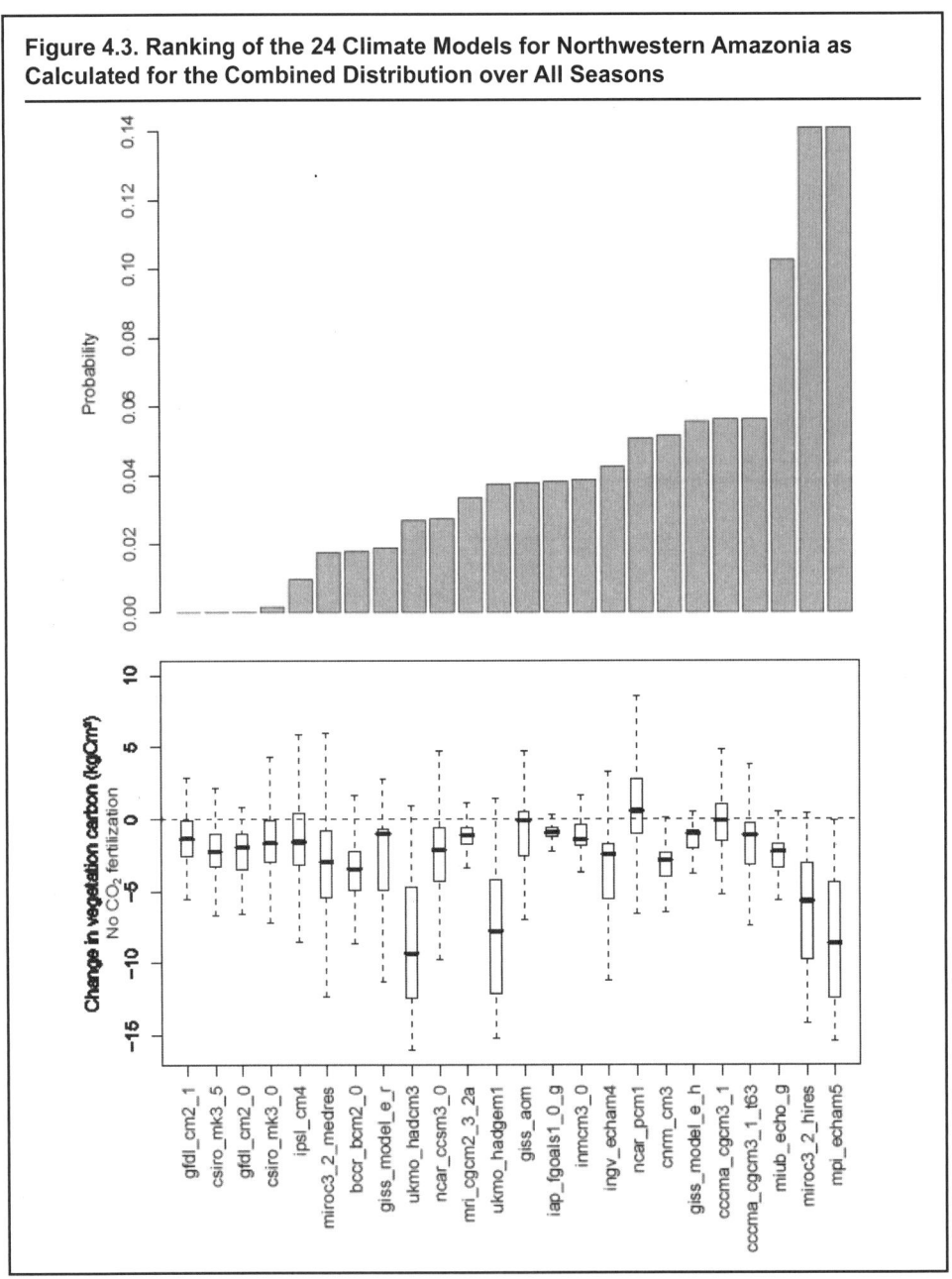

Source: Figure generated for the report by Rammig et al. 2009.
Note: Top panel, models with highest probability are best reproducing mean and distribution of rainfall patterns in this region.

In **Northwestern Amazonia (NWA),** the region with the highest biomass density under current conditions, the range of potential biomass changes is very large. Again, the two highest ranked models project different responses in biomass under the no-CO_2 fertilization scenario. Under the mpi_echam5-climatology, LPJmL projects on average a loss of biomass of ~8 kg C m^{-2} in the grid cells of Northwestern Amazonia. The miroc3_2_hires-climatology leads to a projected decrease on average of ~2 kg C m^{-2}.

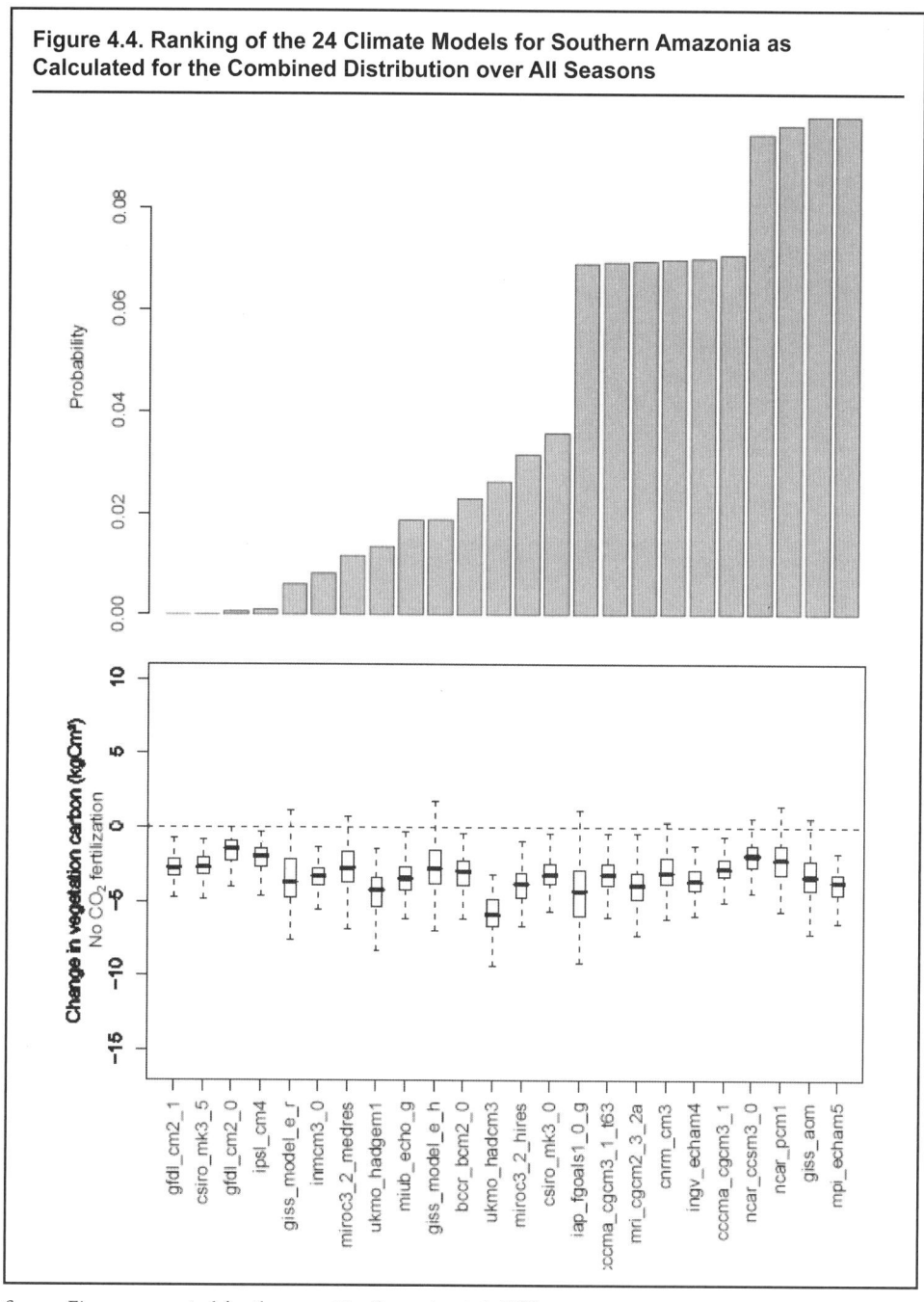

Figure 4.4. Ranking of the 24 Climate Models for Southern Amazonia as Calculated for the Combined Distribution over All Seasons

Source: Figure generated for the report by Rammig et al. 2009.
Note: Top panel, models with highest probability are best reproducing mean and distribution of rainfall patterns in this region.

For **Southern Amazonia (SAz),** LPJmL projects a decrease in vegetation carbon for the four highest ranked climate models (mpi_echam5, giss_aom, ncar_pcm1, ncar_ccsm3_0).

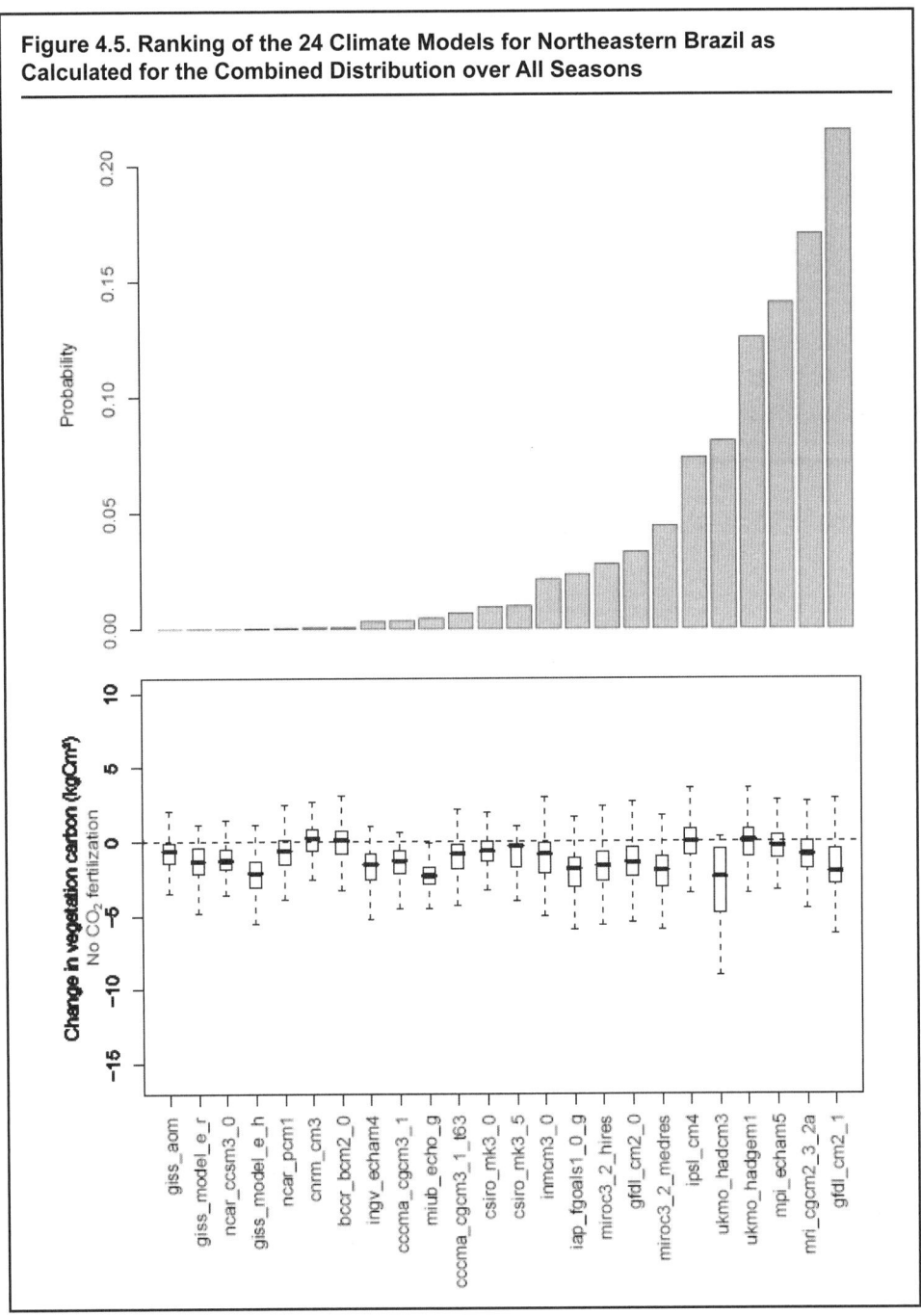

Figure 4.5. Ranking of the 24 Climate Models for Northeastern Brazil as Calculated for the Combined Distribution over All Seasons

Source: Figure generated for the report by Rammig et al. 2009.
Note: Top panel, models with highest probability are best reproducing mean and distribution of rainfall patterns in this region.

In **Northeastern Brazil (NEB),** the gfdl_cm2_1 climate model is highest ranked. Here, LPJmL projects a somewhat lower decrease in vegetation carbon of 0 to -2 kg C m^{-2} for the four best ranked models.

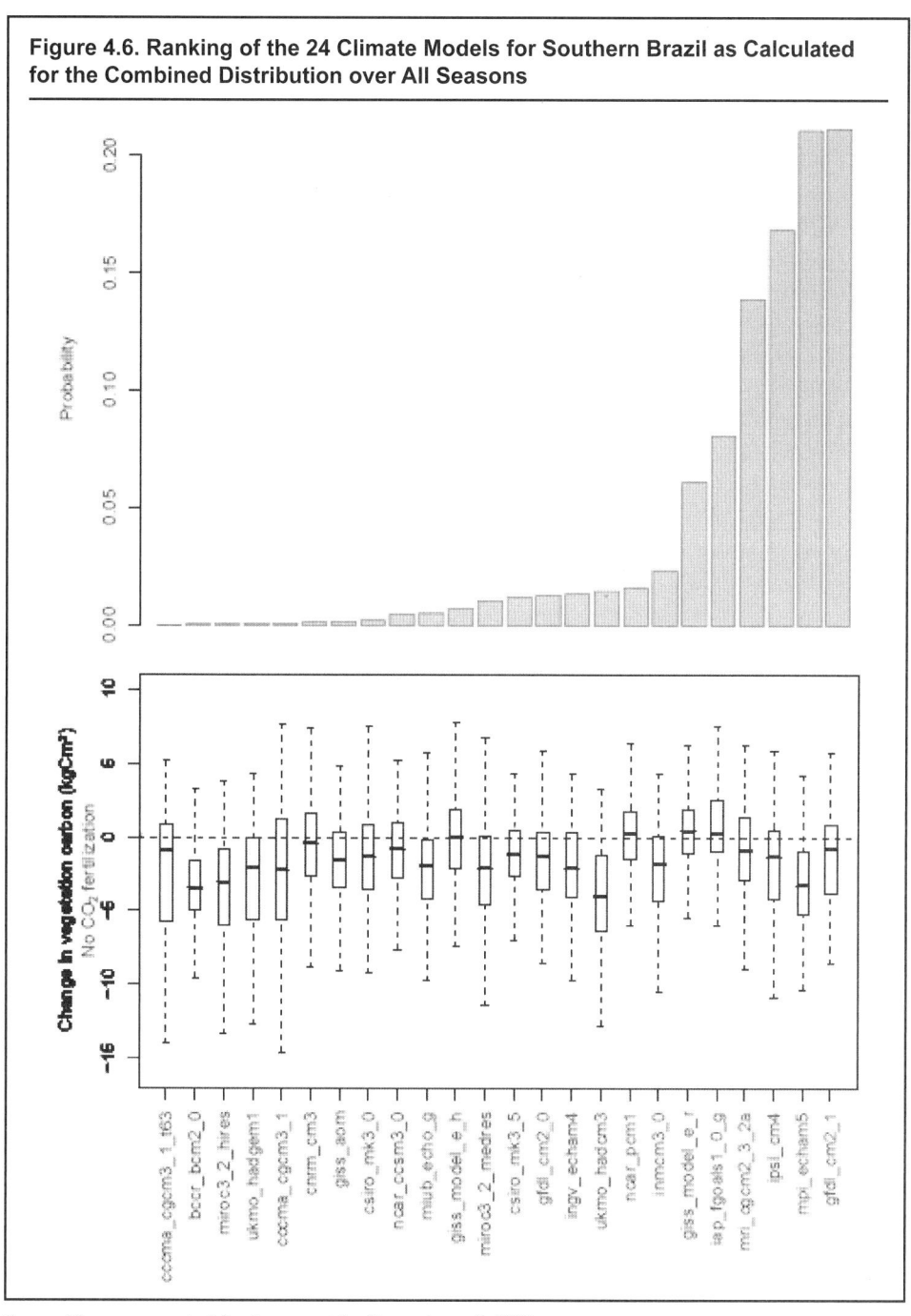

Figure 4.6. Ranking of the 24 Climate Models for Southern Brazil as Calculated for the Combined Distribution over All Seasons

Source: Figure generated for the report by Rammig et al. 2009.
Note: Top panel, models with highest probability are best reproducing mean and distribution of rainfall patterns in this region.

In **Southern Brazil (SB),** vegetation carbon is projected to decrease in most climate scenarios. With the two highest ranked models for this region (mpi_echam5, gfdl_cm2_1), LPJmL projects a "moderate" decrease in biomass.

The plots indicate that the best ranking models consistently predict a reduction in the density of vegetation carbon in all geographical domains; the reductions are largest for those models with the highest ranking.

The vegetation carbon response based on the GCM weightings clearly shows that the changes in vegetation carbon vary strongly among the five regions and among the climate scenarios. The climate models having the highest probability in reproducing annual precipitation cycle differ among the five regions looked at. The vegetation response to the highest-ranked climate models is considerable and often contains a high spatial variability for a climate scenario.

When the CO_2 fertilization effect is included in the estimate of biomass response, the reductions in vegetation carbon are diminished and in some cases an increase in vegetation carbon can be observed. However, as indicated above, the CO_2 fertilization effect in mature forests, under nutrient-limiting conditions such as those prevalent in Amazonian soils, is highly uncertain; therefore, the current information scenarios without CO_2 fertilization should be used as a basis for policy decisions.

The analysis above assumes no CO_2 fertilization. When CO_2 fertilization would be allowed in LPJmL, a number of climate models would result in increases in biomass carbon. However, as indicated before, the extent and limits of CO_2 fertilization are uncertain and subject to current scientific debate (Nowak et al. 2004; Korner et al. 2005; Norby et al. 2005).

Probability Function for Amazon Forest Biomass Change

In this subsection, key outputs of chapters 3 and 4 are combined to derive PDFs and Cumulative Distribution Functions (CDFs) for changes in vegetation carbon. The simulated changes in vegetation carbon from the LPJmL-S1 simulations (shallow roots and the SRES-A1B emission trajectory), as presented in chapter 4, are weighted according to the climate model PDFs shown in figure 3.2. LPJ simulations with and without CO_2-fertilization are considered.

Figures 4.7–4.11 show the derived distribution functions for the five study regions. In each case the inclusion of CO_2 fertilization (red lines) significantly increases the resilience of the forest to CO_2-induced climate change and therefore substantially reduces the risk of forest dieback. This is seen most clearly in the CDF plots (top panels), which show how the probability of a vegetation carbon change below some value x varies with x. Thus, for example, the probability of a reduction in forest carbon greater than 1 kg C m^{-2} (as shown by the vertical dashed line) is about 30 percent in Eastern Amazonia (figure 4.7.) in the absence of CO_2 fertilization (black curve), but is essentially negligible once the default LPJ CO_2-fertilization effects are included (red curve). As mentioned above, there is still an active scientific debate about the likely level of direct CO_2 effects on mature forest ecosystems, with some scientists arguing that nutrient requirements will most likely limit the level of CO_2 fertilization in the 21st century.

On the basis of the PDF of biomass response, the probability of a 25 percent reduction in standing carbon induced by climate change for the different geographical domains has been estimated in the absence of the CO_2 fertilization effect and under the A1B emission trajectory. The assessment estimates a very high probability of dieback in Southern Amazonia (62 percent) and significant probabilities in Northeastern Brazil and Eastern Amazonia. Northwestern Amazonia appears in the analysis to be among the most resilient regions. A more rapid warming may result in more drastic dieback (see table 4.1 below).

Figure 4.7. Cumulative Distribution Functions (Upper Panel) and Probability Density Functions (Lower Panel) for Change in Vegetation Carbon in Eastern Amazonia (from 1970–2000 to 2070–2100)

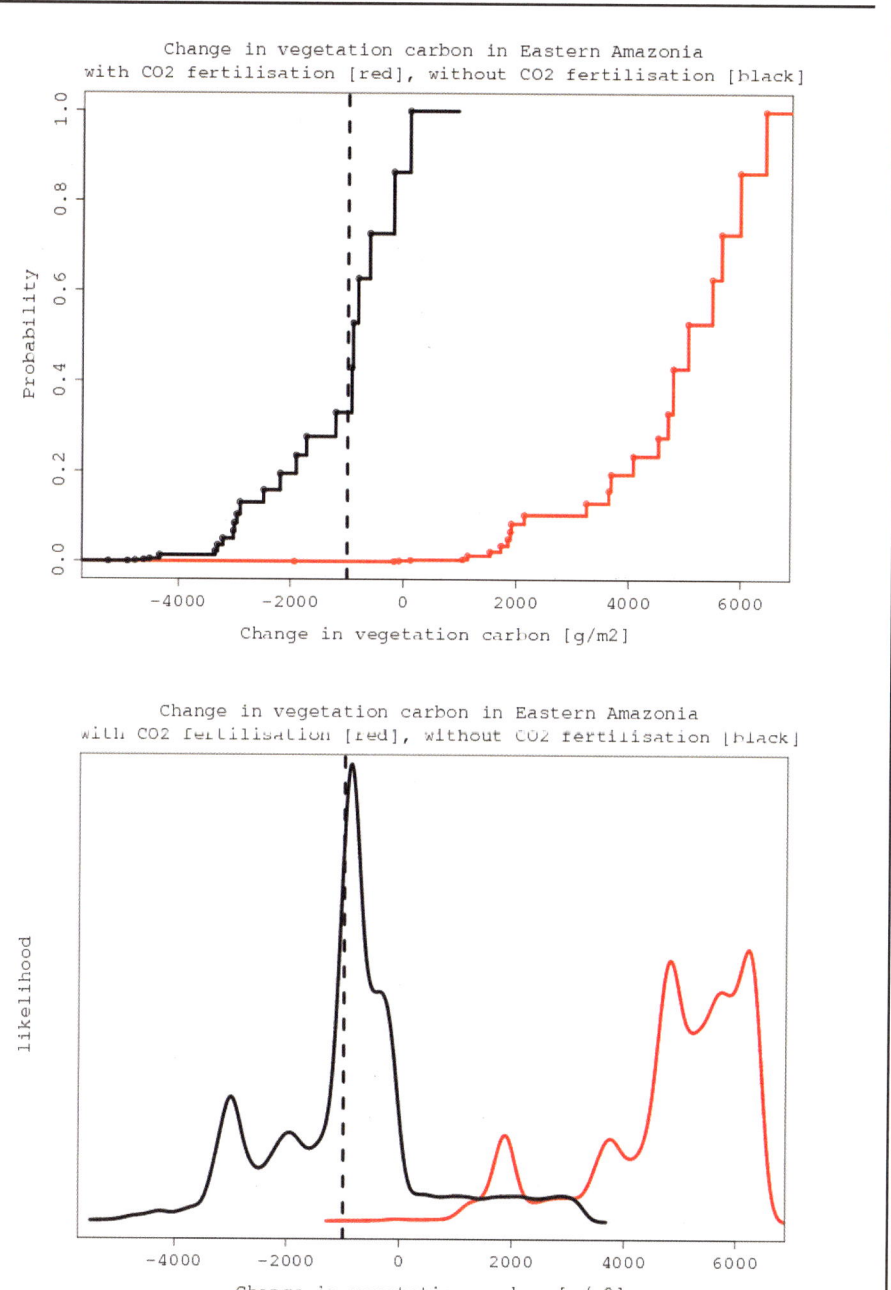

Source: Figure generated for the report by Cox and Jupp 2009.
Note: Produced from LPJmJ-S1 simulations (shallow roots and the SRES-A1B emission trajectory) with CO_2 fertilization (red lines) and without CO_2 fertilization (black). The dotted vertical line indicates a loss of 1 kg C m^{-2}.

Figure 4.8. Cumulative Distribution Functions (Upper Panel) and Probability Density Functions (Lower Panel) for Change in Vegetation Carbon in Northwestern Amazonia (from 1970–2000 to 2070–2100)

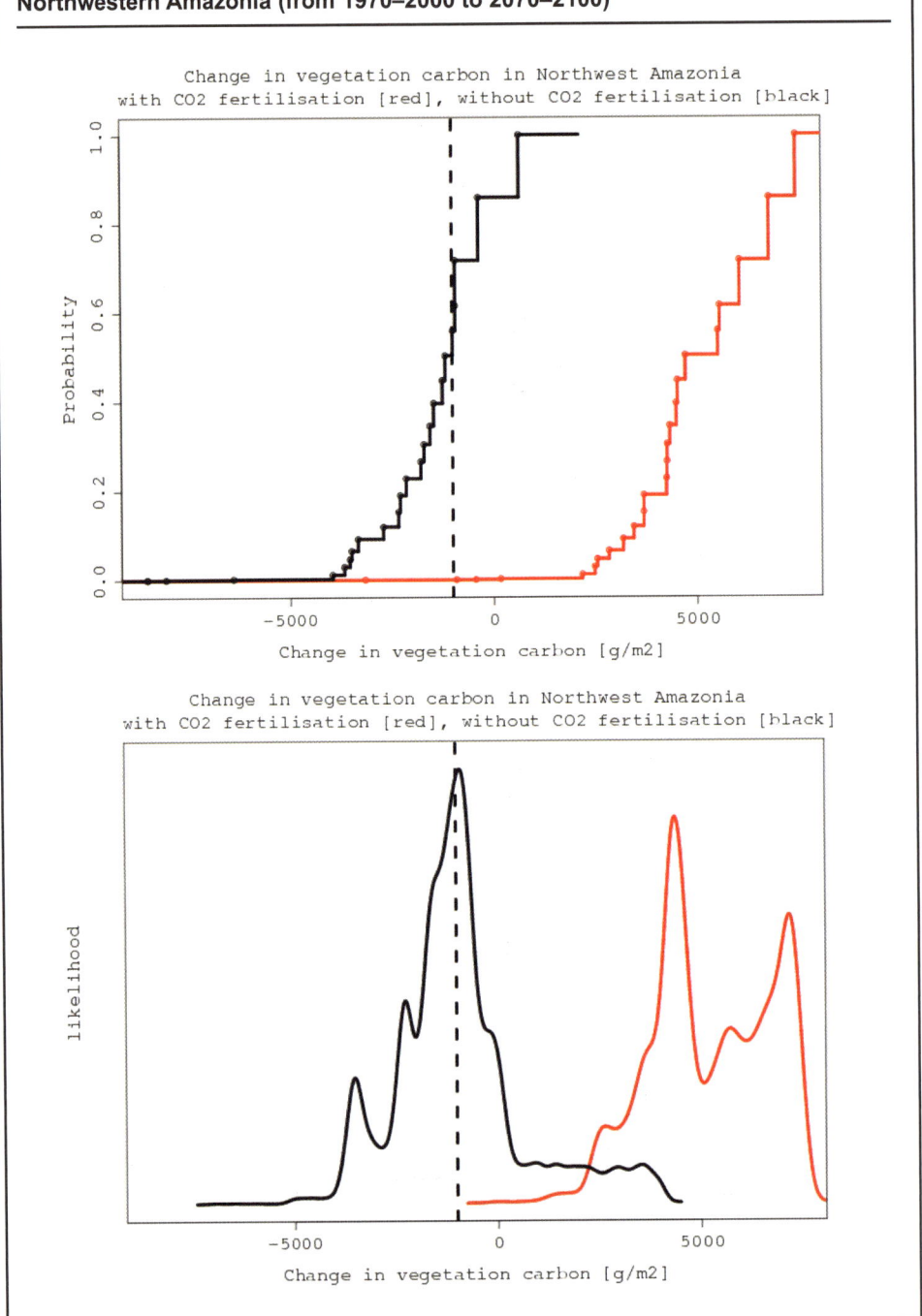

Source: Figure generated for the report by Cox and Jupp 2009.
Note: Produced from LPJmJ-S1 simulations (shallow roots and the SRES-A1B emission trajectory) with CO_2 fertilization (red lines) and without CO_2 fertilization (black). The dotted vertical line indicates a loss of 1 kg C m^{-2}.

Figure 4.9. Cumulative Distribution Functions (Upper Panel) and Probability Density Functions (Lower Panel) for Change in Vegetation Carbon in Southern Amazonia (from 1970–2000 to 2070–2100)

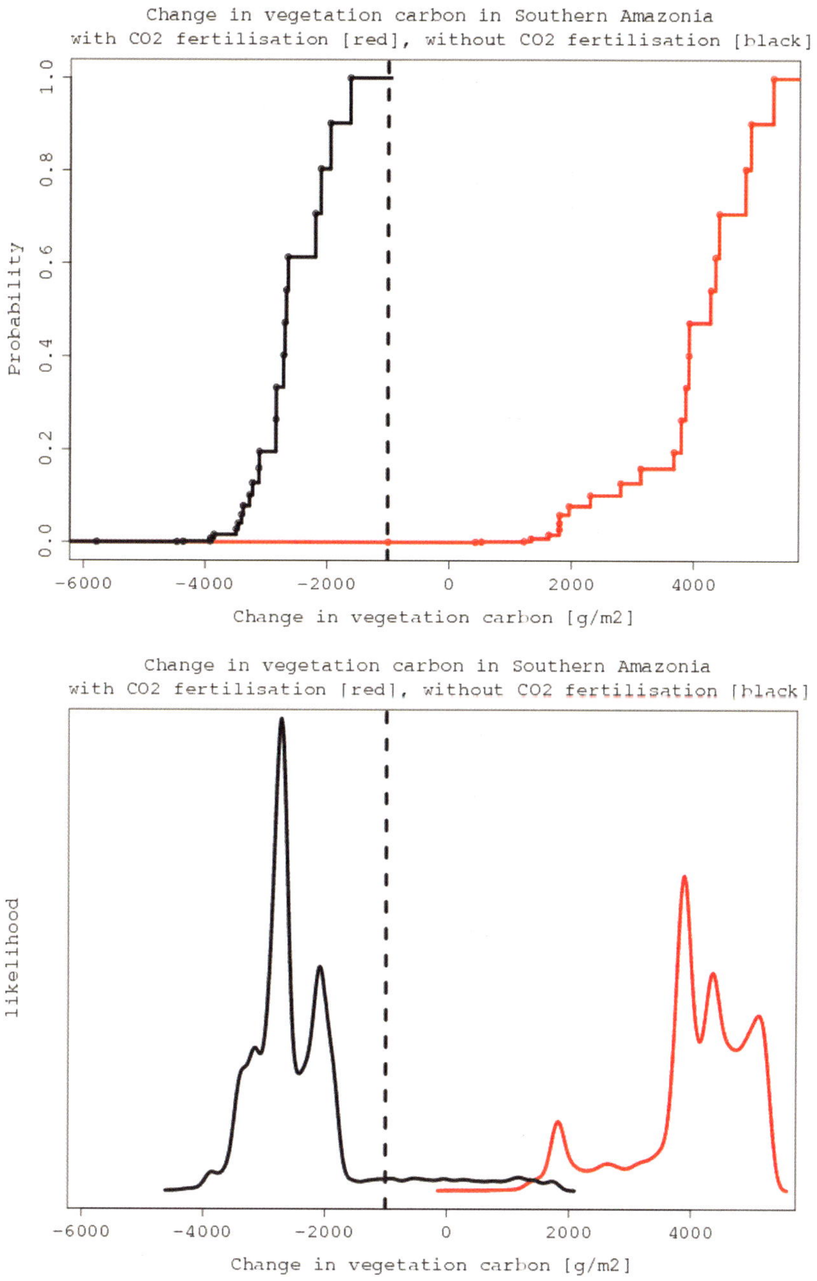

Source: Figure generated for the report by Cox and Jupp 2009.
Note: Produced from LPJmJ-S1 simulations (shallow roots and the SRES-A1B emission trajectory) with CO_2 fertilization (red lines) and without CO_2 fertilization (black). The dotted vertical line indicates a loss of 1 kg C m^{-2}.

Figure 4.10. Cumulative Distribution Functions (Upper Panel) and Probability Density Functions (Lower Panel) for Change in Vegetation Carbon in Northeastern Brazil (from 1970–2000 to 2070–2100)

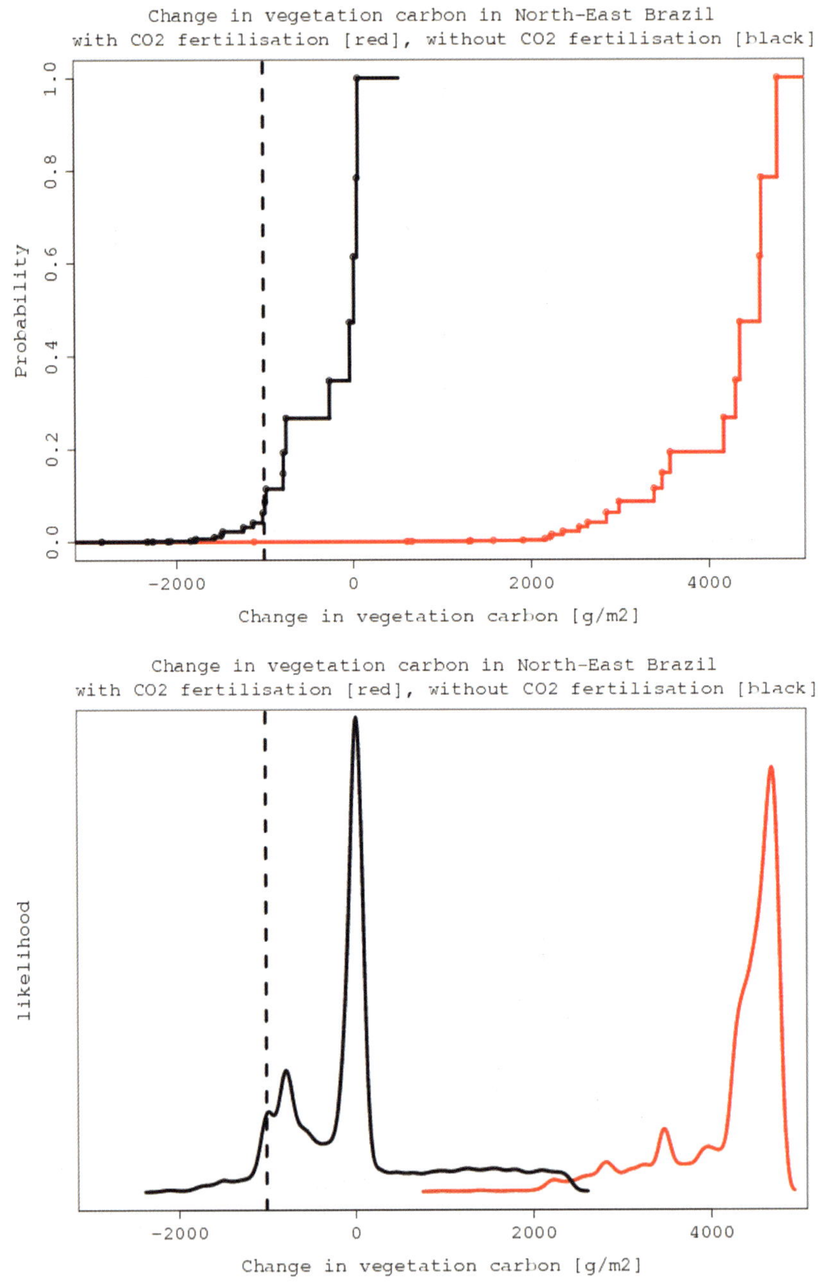

Source: Figure generated for the report by Cox and Jupp 2009.
Note: Produced from LPJmJ-S1 simulations (shallow roots and the SRES-A1B emission trajectory) with CO_2 fertilization (red lines) and without CO_2 fertilization (black). The dotted vertical line indicates a loss of 1 kg C m^{-2}.

Figure 4.11. Cumulative Distribution Functions (Upper Panel) and Probability Density Functions (Lower Panel) for Change in Vegetation Carbon in Southern Brazil (from 1970–2000 to 2070–2100)

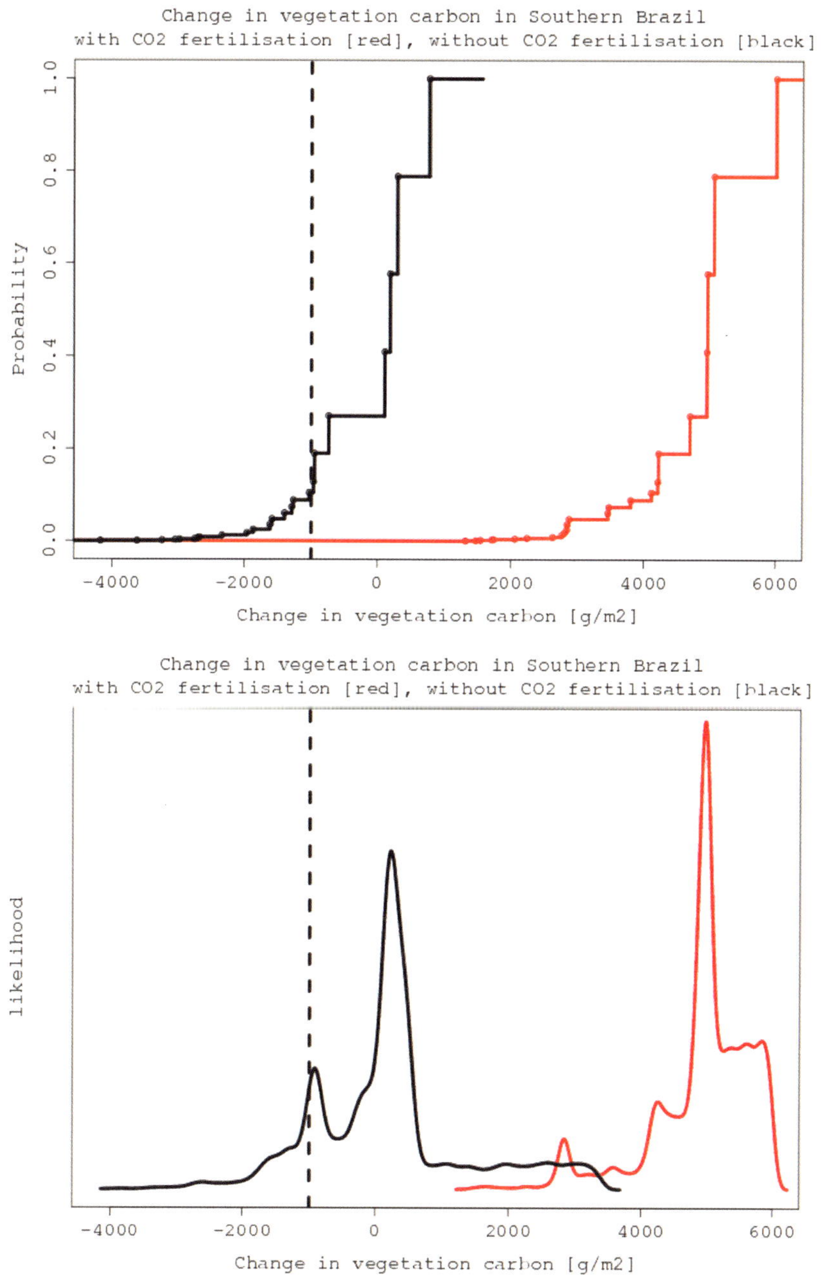

Source: Figure generated for the report by Cox and Jupp 2009.
Note: Produced from LPJmJ-S1 simulations (shallow roots and the SRES-A1B emission trajectory) with CO_2 fertilization (red lines) and without CO_2 fertilization (black). The dotted vertical line indicates a loss of 1 kg C m^{-2}.

Table 4.1. Assessment of the Risk of Amazon Dieback, Defined as 25% Loss of Vegetation Carbon (from 1970–2000 to 2070–2100) for the Five Geographical Domains Using LPJmL-S1 Simulations (Shallow Roots and the SRES-A1B Emission Trajectory) Without CO_2 Fertilization

Geographical domain	Average veg. carbon (kgC/m^2)	Probability of dieback w/o CO_2 fertilization in %
Eastern Amazonia	12	8
Northwestern Amazonia	15	3
Southern Amazonia	10	62
Northeastern Brazil	3	19
Southern Brazil	10	2

Source: Table generated for the report.

It is concluded that direct CO_2 effects at ecosystem level are the key unknown in assessing the risk of Amazon forest dieback under 21st century climate change.[4] Reducing this uncertainty, using a combination of estimates of the Amazonian carbon sink (Phillips et al. 2009) and trends in river runoff (Gedney et al. 2006), is therefore a priority for follow-on study. Drastic differences between the total impacts of climate change caused by CO_2 and other climate forcing agents (e.g., increases in methane or reductions in sulphate aerosols) have implications for international climate policy, which currently treats all radiative forcings as equally damaging. In contrast, this study shows clearly that the risk of Amazon forest dieback is many times greater if climate change arises from agents other than CO_2.

Simulation of Sensitivity to CO_2 and Rooting Depth

As seen above, the likelihood and extent of CO_2 fertilization and its effects at ecosystem level are critical to assess the prospects for dieback. Root depth also affects the resilience of biomass, in particular its ability to withstand droughts. To evaluate the effects of CO_2 fertilization and different rooting depths, LPJmL was run for four scenarios (table 4.2). Scenarios S1 and S2 are run using the standard SRES-A2 emission trajectory and include the effects of climate and increasing atmospheric CO_2 conditions. Under scenarios S3 and S4, the effect of CO_2 has been removed by using constant 2000 CO_2 conditions.

Scientific studies show that the effects of elevated CO_2 on plant growth are dependent on (i) the species considered, (ii) the growth stage of the species, (iii) its photosynthetic characteristics, as well as (iv) the management regime, such as water, nitrogen (N) applications for crops and availability of micro-nutrients (Jablonski et al. 2002; Kimball et al. 2002; Norby et al. 2003; Ainsworth and Long, 2005).[5]

For trees, the measured change in biomass (i.e. growth) in young and rapidly growing parcels at 550 ppm CO_2 can be in the 0–30 percent range, with the higher values reported in younger trees.[6] For mature, natural forests there was no or only little response to elevated CO_2 observed (Nowak et al. 2004; Korner et al. 2005; Norby et al. 2005).[7] Norby et al. (2005) found a mean tree net primary production (NPP) response of 23 percent in young tree stands; however, in mature tree stands Korner et al. (2005) reported no CO_2 stimulation. In addition, it is important to note that the measured NPP response (if so) is not providing any information on net changes in carbon stock (Korner et al.

2007). Thus, the effect of CO_2 stimulation at ecosystem level on mature tropical forest of the type prevalent in the Amazon basin has yet to be ascertained, particularly under soil nutrient constraints.

For assessment of the effects of rooting depth, two different model settings of LPJmL were also used. In the standard setting (scenarios S1 and S3), the soil is differentiated in two layers: the upper layer contains 50 cm of soil and the lower layer 150 cm. 85 percent of the roots of evergreen PFTs are located in the upper and 15 percent in the lower soil layer. Raingreen trees are assumed to have 60 percent of their roots in the upper and 40 percent in the lower soil layer. As a second scenario, simulating "deep roots" (scenarios S2 and S4), a rooting depth of 8 m, with a soil profile of 50 cm of soil in the upper, and 750 cm in the lower layer was assumed. It was also assumed that evergreen trees have deeper roots with only 55 percent of their roots in the upper and 45 percent in the lower layer. Raingreen trees were assumed to have 85 percent of their roots in the upper and only 15 percent of the roots in the lower layer.

Table 4.2. Simulation Experiments Conducted with LPJmL to Investigate the Role of CO_2 and Deep Roots

Scenario	Description	
S1	Climate and CO_2 effects	Shallow roots
S2	Climate and CO_2 effects	Deep roots
S3	Climate effects only	Shallow roots
S4	Climate effects only	Deep roots

Source: Table generated for the report by Rammig et al. 2009.
Note: See text for detailed description of the experiments.

LPJmL simulates the coupled terrestrial carbon and water cycles, which are linked through vegetation with roots growing to a certain depth in the soil layer. However, only a few observations on rooting depth and distribution exist. Recent investigations have outlined the importance of deep roots to maintain a close canopy during the dry season, where they are needed to obtain sufficient water supply from deep soil water. Standard LPJmL simulations assume a soil depth of 2 m but root systems of up to 18 m deep have been found in Northeastern Pará (Nepstad et al. 1994; Kleidon and Heimann 2000). In order to better understand the associated uncertainty, a simulation experiment was conducted in which the effect of deep and shallow roots was tested (table 4.3).

In a similar way, the effects of rising atmospheric CO_2 concentrations on vegetation are uncertain. CO_2 plays a major role as a limiting resource for carbon assimilation by plants (Farquhar et al. 1980). Several small-scale and chamber experiments have shown an enhancement of photosynthesis in C_3-plants under elevated CO_2 concentrations, leading to increased NPP (Curtis and Wang 1998; Norby et al. 1999). However, the long-term effects on real ecosystems are unclear (Norby et al. 1999). Large-scale ecosystem models such as LPJmL generally suggest a substantial impact of CO_2 on NPP (Cramer et al. 2001). Measurements from large-scale free-air CO_2 enrichment (FACE) experiments in temperate forests (Norby et al. 2005) have been compared to LPJ model simulations, and have shown that the model reproduced the overall response of forest productivity to elevated CO_2 (Hickler et al. 2008). In the model assumptions, elevated CO_2 concentra-

Table 4.3. Percentage of Forest Cover of the Classified Vegetation Types in the Amazon Basin under HadCM3-A2 Scenario

	Forest cover (%)					
	1991–2000					
Scenario	Tropical	Deciduous	Open forest	Woodland	Shrubland	Savanna
S1	45.5	46.5	0.6	1.5	0.2	5.8
S2	75.3	16.8	0.3	1.5	0.2	5.8
S3	45.5	46.5	0.6	1.5	0.2	5.8
S4	75.3	16.8	0.3	1.5	0.2	5.8
	2041–2050					
	Tropical	Deciduous	Open forest	Woodland	Shrubland	Savanna
S1	32.0	60.8	0.2	1.2	0.1	5.8
S2	69.4	23.1	0.2	1.5	0.1	5.8
S3	25.3	27.8	20.1	5.0	15.9	5.8
S4	45.5	20.1	3.4	13.7	11.4	5.8
	2091–2100					
	Tropical	Deciduous	Open forest	Woodland	Shrubland	Savanna
S1	15.2	55.7	2.7	8.2	10.3	7.7
S2	36.9	38.9	2.9	9.0	5.9	6.4
S3	0.5	5.8	6.9	1.9	67.6	17.3
S4	0.8	2.5	2.8	3.1	74.7	16.1

Source: Table generated for the report by Rammig et al. 2009.
Note: Results for the factorial experiments, in which the effects of climate and CO_2 and the effects of shallow and deep rooting trees were tested.

tions reduce the negative effects of drought on plant growth (Gerten et al. 2005), which increase plant productivity. In the current assessment, the effects of CO_2 were tested using a constant CO_2 scenario (table 4.3) for a model that predicts severe rainfall reduction and Amazon dieback (HadCM3).

Varying rooting depth and CO_2 effects produced strong effects on the degradation of Amazon forests under climate change (table 4.3). Under current conditions, the vegetation of the Amazon basin is dominated by "tropical" and "mixed" deciduous forests. Assuming a forest with shallow rooting trees (scenario S1), 46 percent of the Amazon basin is covered with tropical forests (HadCM3-A2 scenario). With the assumption of deeper rooting trees (scenario S2), a 75 percent cover with tropical forests is projected for current climate conditions, due to better accessibility of trees to deep soil water.

With increasing atmospheric CO_2 concentrations (SRES-A2), the degradation of tropical forests is moderate until the middle of the 21st century, with 32 percent and 69 percent of the Amazon basin still covered by tropical forests in the S1 and S2 scenarios, respectively (figure 4.12). Removing the drought-buffering and growth-stimulating effects of CO_2 from the simulations leads to a different picture. In the S3 scenario, about half of the former "mixed" forests are degraded to "open" forests, containing lower amounts of biomass. Shrublands increase by 15 percent. In the deep roots scenario, woodland and shrubland increase by 10 percent.

By the end of the century, under the assumed climate and CO_2 effects, only 37 percent of tropical forests remain in the S2 Scenario and 15 percent in the S1 Scenario. In the

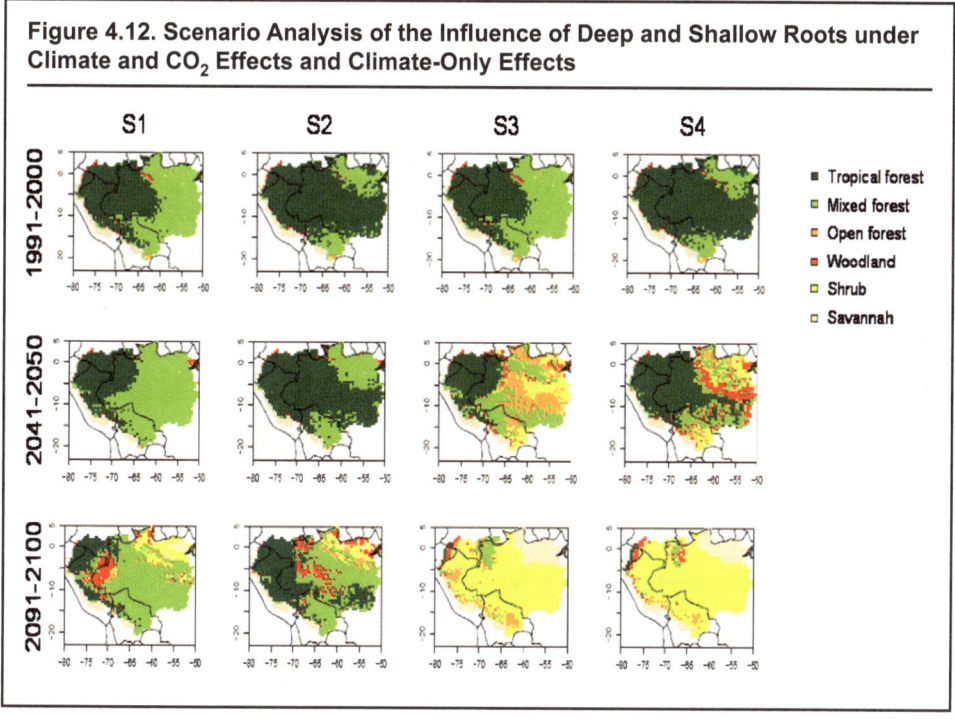

Figure 4.12. Scenario Analysis of the Influence of Deep and Shallow Roots under Climate and CO₂ Effects and Climate-Only Effects

Source: Figure generated for the report by Rammig et al. 2009.
Note: Simulation results for the HadCM3-A2 climate scenario.

extreme case of the S3 Scenario, tropical forests disappear. The vegetation in the basin is converted to 68 percent shrubland and 17 percent savanna-like vegetation. The remaining 15 percent consists of mixed and open forests. Similar patterns are found in the S4 Scenario.

These results lead to the conclusion that the extent of a potential Amazon forest dieback is highly dependent on the model assumptions made on vegetation structure and response to environmental factors. The evaluation of the two assumptions on (1) the effects of CO_2 buffering on drought and (2) different rooting depth showed that an estimate of the strength of a potential dieback and the direction of forest degradation is strongly dependent on these assumptions.

Evidence from field measurements for these processes is lacking and estimates on the potential effects of CO_2 or on rooting depth distribution over the Amazon basin are not yet available. These crucial aspects for the ecosystem's resilience would certainly have to be part of any follow-up analysis. A substantial impact of CO_2 fertilization in mature stands of tropical forest under nutrient limited conditions is uncertain and should not be used as the basis for policy decisions.

Changes in Transpiration

Another aspect of vegetation changes in the Amazon basin is the contribution of forest to the convective precipitation that plays an important role in the region for water supply (Malhi et al. 2008). Vegetation in Amazonia is dominated by tropical and mixed

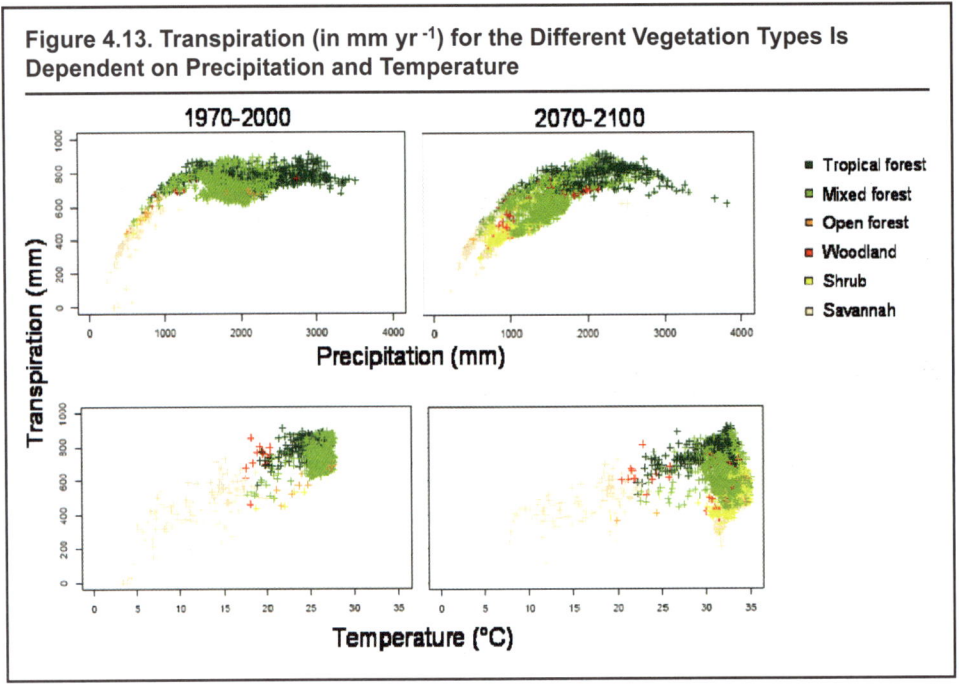

Figure 4.13. Transpiration (in mm yr⁻¹) for the Different Vegetation Types Is Dependent on Precipitation and Temperature

Source: Figure generated for the report by Rammig et al. 2009.
Note: Under future conditions, with changing precipitation and temperature patterns and shifts in vegetation, transpiration changes. Results from LPJmL simulations with HadCM3-A2 climatology.

forests that maintain high transpiration rates of ~850 mm yr⁻¹ (~2.3 mm d¹) under current climate conditions, with annual precipitation of 1200–3000 mm and temperatures of 20–28°C. Figure 4.13 illustrates that tropical forests have a lower plasticity when it comes to accommodation to drier conditions. They remain only in areas with still high precipitation and maintain high transpiration rates. Mixed forests are able to adapt to drier conditions and their transpiration rates decrease under drier conditions and with higher temperatures. However, the ecosystem shifts toward degraded forest types with less forest cover and lower biomass, such as woodland and shrubs. These types transpire less and therefore reduce the contribution to convective precipitation.

Mechanisms of Potential Amazon Dieback

The mechanism of potential forest dieback is illustrated best by the development of forest and grass cover over time in regions with severe drought simulated by the climate model. This is illustrated for one example grid cell located in the northeastern part of Amazonia, at 51.25°W and 0.25°S (figure 4.14), using one of the models that predict rainfall reductions.

The overall driver of the climate-caused collapse of forest in the northeastern part of Amazonia under the HadCM3-A2 climatology is a general decrease in mean precipitation accompanied by several extreme drought events starting after 2050. High transpiration rates first lead to a depletion of soil water in the upper and then the lower soil layer, causing high water stress in the plants. This is followed by a strong reduction in NPP and quickly increasing mortality of trees.

Figure 4.14. Vegetation Change at Local Scale in a Grid Cell in Northeastern Amazonia

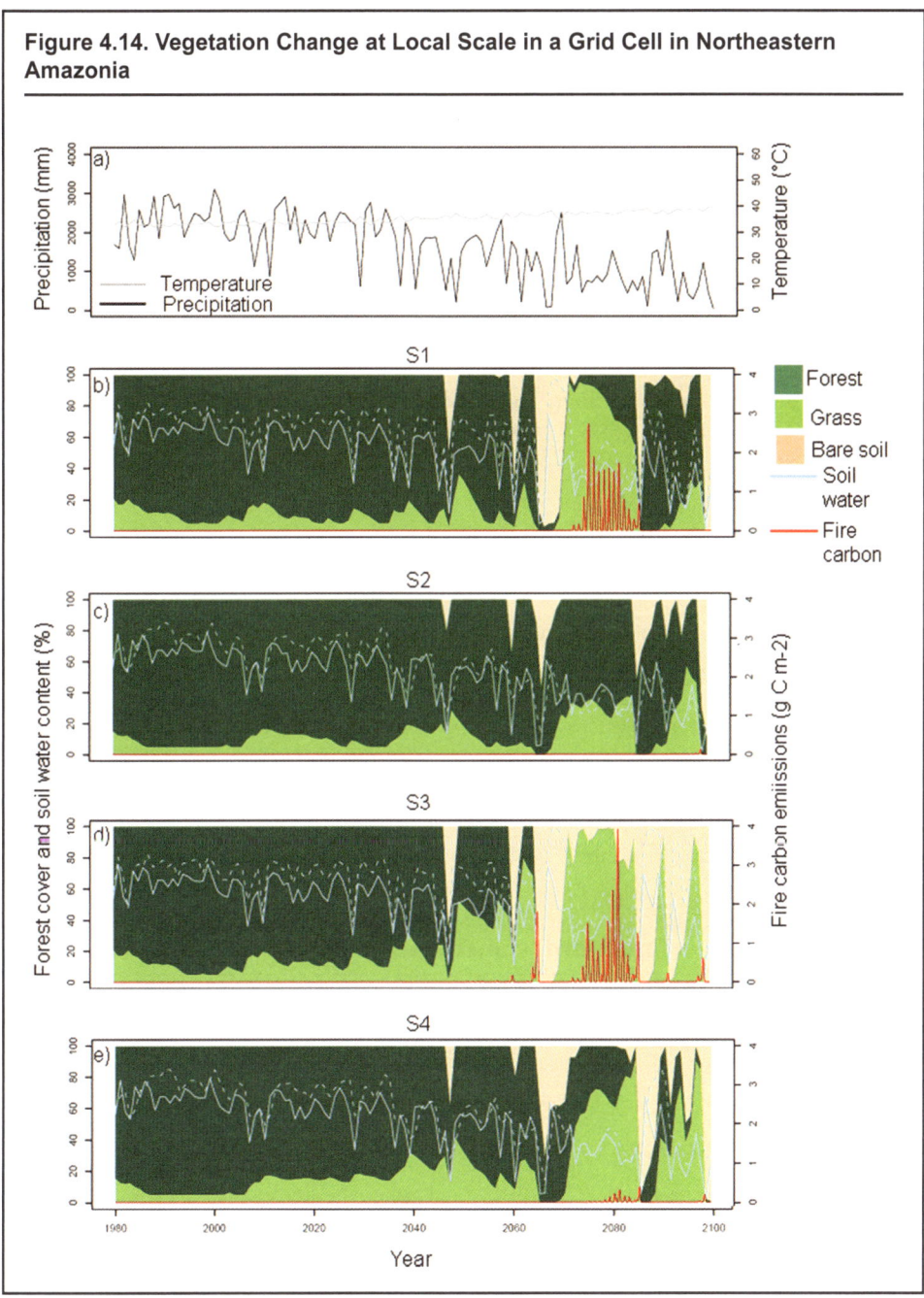

Source: Figure generated for the report by Rammig et al. 2009.
Note: Figure shows (a) Mean annual temperature and sum of annual precipitation from 1980 to 2100 from the HadCM3-A2 climate scenario; (b-e) Factorial analysis of the influence of deep/shallow roots and simulations with influence of climate+CO2 and climate change effects only (Scenarios S1-S4). The light blue line is soil water content as a fraction of soil water holding capacity (in percent) for upper (plain line) and lower (dashed line) soil levels.

In this particular example, forest cover recovers during the following years back to 80–95 percent. This can be explained by a quick response of woody-vegetation re-growth. The response of the vegetation may be severely overestimated in LPJmL since no soil erosion or other degradation is assumed to take place. The structure of the forest changes, with a result of more grass vegetation than before the drought events. In the case of the S1 and S3 scenarios, the high amount of these highly flammable grasses triggers frequent fire events. These effects are stronger under the climate-effects-only assumption. The system may at this point change to a fire-driven ecosystem state that can be characterized as savanna.

Changes in Lightning-Caused Wildfires

Fires are rare events under undisturbed conditions in tropical forest ecosystems. They have been observed historically either as part of small-scale slash-and-burn activity (Kauffman and Uhl 1990), or due to lightning-caused ignitions in occasional drought years (Cochrane and Laurance 2008). Thus, besides physiology-driven growth and mortality responses, another important indicator for the effects of climate change on the Amazon rainforest could be changes in the occurrence of actual fires or fire danger.

Climatic fire danger, based on LPJmL-SPITFIRE, was analyzed. The fire danger index is projected to increase significantly under the dry and hot HadCM3-A2 Scenario. The interannual variability increases remarkably after 2060, when the projected precipitation starts to decline in the HadCM3 model. Under the wetter scenario of the MRI CGCM 2.3.2a, the climatic fire risk remains variable but relatively constant over the two centuries (figure 4.15).

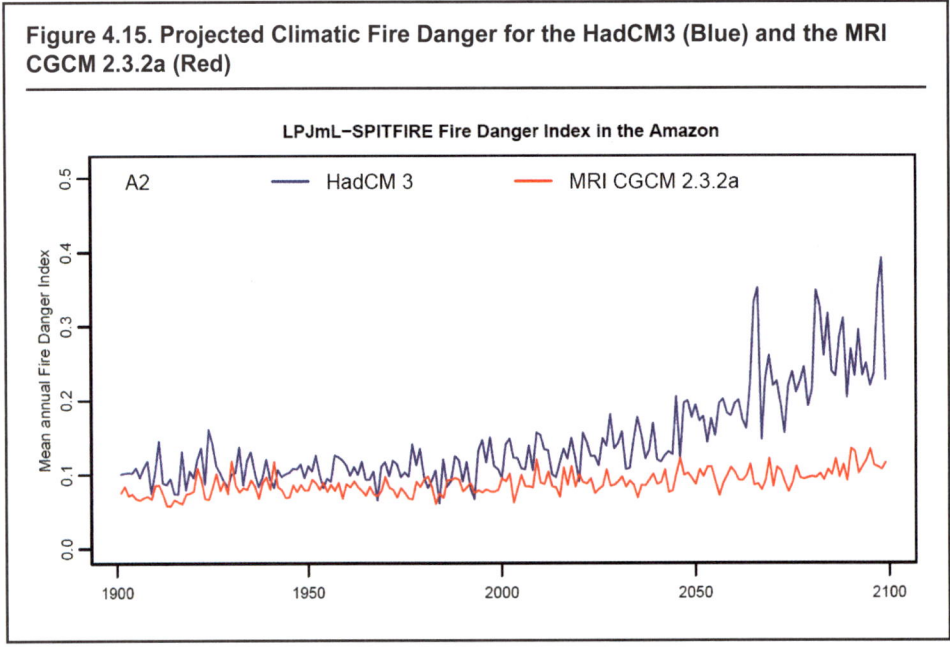

Figure 4.15. Projected Climatic Fire Danger for the HadCM3 (Blue) and the MRI CGCM 2.3.2a (Red)

Source: Figure generated for the report by Rammig et al. 2009.
Note: The projection is made under the SRES-A2 Scenario simulated by LPJmL-SPITFIRE for the 20th and 21st century.

Using the HadCM3-A2 Scenario, the simulated climatic fire danger is still low under current climate conditions, but already higher than under the MRI CGCM. Changed climatic conditions by the end of the 21st century lead to an increased fire danger, with very high danger levels in Northeastern Amazonia in the HadCM3-A2 Scenario. In the wet scenario (MRI CGCM 2.3.2a), climatic fire danger increases from very low to low fire danger levels, mainly in the southeast of the basin.

The elevated fire danger index is not automatically leading to increased fire frequency. Fires can start after lightning events only if sufficient fuel load is available. Thus, after a significant increase of flammable grasses, e.g., as a result from drought-induced forest degradation (see figure 4.16), increases in climatic fire danger in the Northeastern Amazon lead to an increase in burned area, thus the fire-related carbon emission (figure 4.17) in the HadCM3-A2 scenario. Low climatic fire danger levels do not allow the development of sufficient surface energy which could sustain burning. Therefore, no carbon emissions are simulated under the wet MRI-CGCM 2.3.2a climate scenario.

Figure 4.16. Simulated Climatic Fire Danger under the MRI CGCM 2.3.2a (Top) and the HadCM3 (Bottom) Climate Scenario Using the SRES A2 Emission Scenario

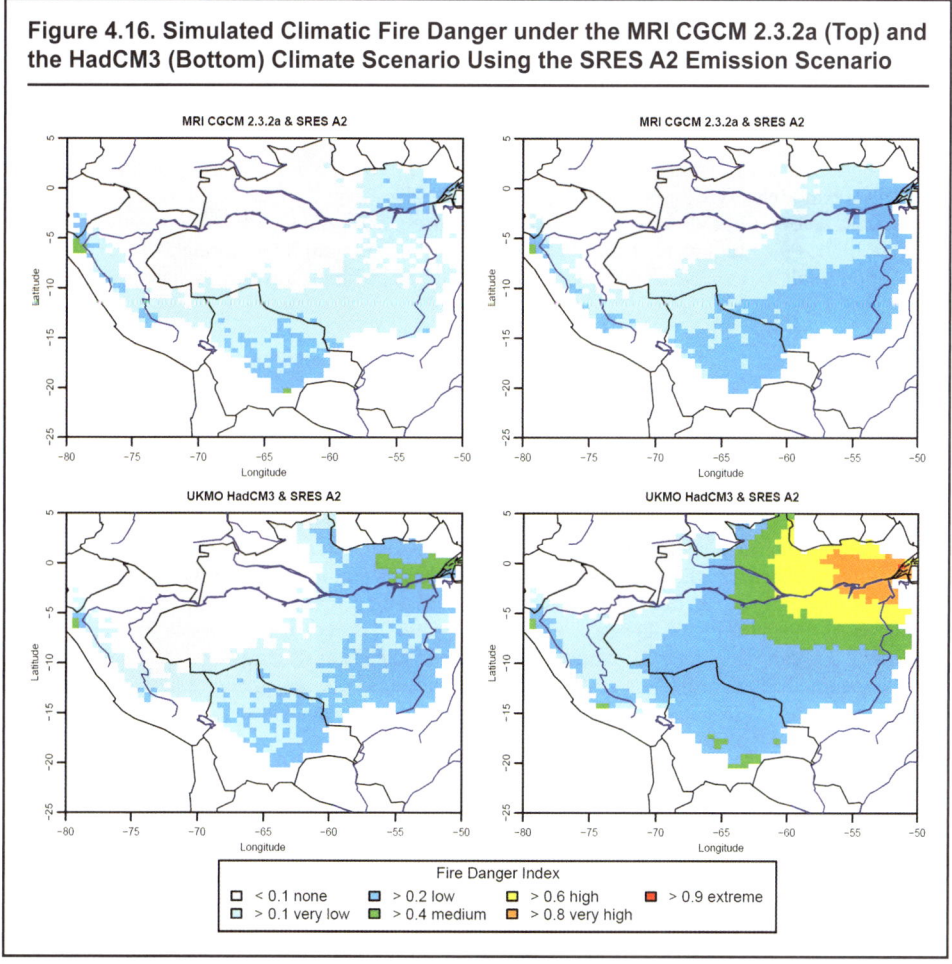

Source: Figure generated for the report by Rammig et al. 2009.
Note: Top-left and bottom-left maps show contemporary average (1970–2000); top-right and bottom-right maps show the average over 2070–2100.

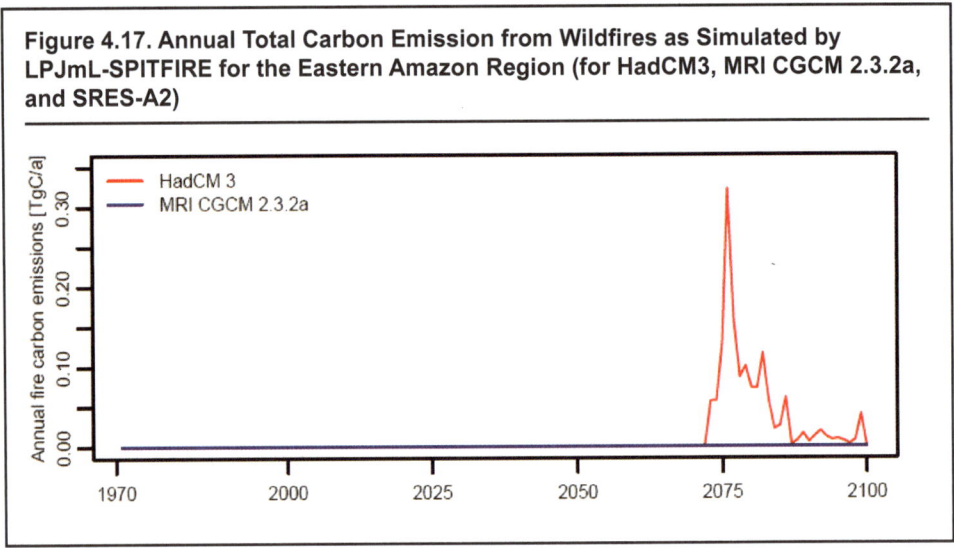

Figure 4.17. Annual Total Carbon Emission from Wildfires as Simulated by LPJmL-SPITFIRE for the Eastern Amazon Region (for HadCM3, MRI CGCM 2.3.2a, and SRES-A2)

Source: Figure generated for the report by Rammig et al. 2009.

Notes

1. For this purpose, dynamic global vegetation models have been developed (Prentice et al. 2007). Several such models exist, based on different conceptualizations, such as TRIFFID (Cox 2001), IBIS (Foley et al. 1996), LPJmL (Sitch et al. 2003; Bondeau et al. 2007) and ORCHIDEE (Essery et al. 2001; Krinner et al. 2005; Hughes et al. 2006). All these models are simplified enough to be generic for application to all major upland ecosystems of the planet, but they can also be used regionally for specific purposes.

2. LPJmL has been evaluated using observations in a large number of studies and against many types of observations, on the global (Gerten et al. 2004) and regional scales for boreal forests (Lucht et al. 2002), and tropical ecosystems (Cowling and Shin 2006).

3. In any grid cell, the simulation is driven by the input of monthly climatology, annual ambient CO_2 and static soil properties.

4. While physiological aspects of CO_2 stimulation at leaf level are well-researched, a major challenge is to upscale CO_2 leaf-level effects to community- or ecosystem-level, which is a critical assumption in vegetation modeling. Korner (2004, 2006) for example shows that with regard to CO_2 fertilization, it is crucially important to clearly distinguish a possible stimulation of growth or NPP from any change in the carbon pool size of the ecosystem. Enhanced growth or NPP, should it occur at ecosystem level, does not translate in a change in pool size by simple terms neither in a change of the ecosystem's resilience. This in turn means that even if there was a CO_2 stimulation of growth, this cannot be taken as evidence of a change in carbon storage or increased resilience of the ecosystem. To the contrary, should there be a stimulation of fast growing, low wood density tree species or lianas (see discussion in Korner 2004), the ecosystem's carbon stock would decline despite greater NPP or growth.

5. Compared to current atmospheric CO_2 concentrations, crop yields, for example, increase at 550 ppm CO_2 in the range of 10–20 percent for C3 crops (e.g., wheat, barley, potatoes) and 0–10 percent for C4 crops such as maize (Ainsworth et al. 2004; Gifford, 2004; Long et al. 2004).

6. However, from such data landscape wide C-stocking cannot be inferred, unless the associated tree life history (life span, turnover etc.) is known.

7. For commercial forestry this would mean that slow-growing trees may respond little to elevated CO_2 (e.g., Vanhatalo et al. 2003), while fast-growing trees would possibly do so more strongly (Calfapietra et al. 2003; Liberloo et al. 2005; Wittig et al. 2005).

CHAPTER 5

Interplay of Climate Impacts and Deforestation in the Amazon

Regional Land Use as a Driver in the Stability of the Amazon Rainforest

The previous analysis was made on the basis of no land use change and in response to climate-induced changes in rainfall through a dynamic (LPJmL) vegetation model. However, regional land use changes, such as deforestation, biomass burning and forest fragmentation, affect local and regional climate and may compound the effects of global climate change on the stability of the Amazon rainforest by redefining bioclimatic conditions and thus the biome-equilibrium state of the basin.

For example, field observations (e.g., Gash and Nobre 1997) and numerical studies (e.g., Sampaio et al. 2007, Nobre et al. 1991) have revealed that large-scale deforestation in Amazonia could alter the regional climate significantly. Evapotranspiration and precipitation are reduced and soil temperature is increased where there is no forest canopy. That effect might lead to a biome shift toward "savanna" of portions of the tropical forest domain.[1]

A coupled climate-vegetation model (CPTEC-CPVM, described later in the report) is used to estimate these combined effects.[2] The advantage of being coupled with a climate model is that the feedbacks of vegetation change to the climate can be investigated. Thus, the model can simulate biome distribution (one biome per grid cell) based on bioclimatic limits as these are affected by climate change.

The current vegetation in the Amazon basin, including its deep root system, is efficient in recycling water vapor, which may be an important mechanism for the forest's maintenance and contributes to the overall water cycle in the region. Thus, deforestation and forest fragmentation can alter the hydrological cycle and cause other impacts as well. For instance, in the event of severe droughts the forest can become highly susceptible to fires due to soil water deficits (Nepstad et al. 2001). Soares-Filho et al. (2006) have shown that if the current high deforestation rates are to continue into the future, about 40 percent of the Amazonian tropical forests will have disappeared by 2050. In principle, deforestation and global warming acting synergistically could lead to drastic biome changes in Amazonia.

In this chapter, analyses are made to quantify how deforestation and climate change may combine to affect the distribution of the Amazon ecosystem. To this end, simulations of climate (temperature, precipitation) and vegetation change in tropical South America were performed in the five selected geographical domains. The simulations account for land use change, global warming, and vegetation fires. Changes in land use consider deforestation scenarios of 0 percent, 20 percent, 50 percent, and 100 percent, with and without fires, under scenarios B1 and A2 from IPCC AR4.

In addition, results of the use of the CPTEC land vegetation model (CPTEC-PVM) with the Earth Simulator MRI-GCM outputs, with particular focus on the regional impacts in the five selected domains and the La Plata basin, are also reported.

It is important to mention that Brazil has announced a plan to reduce deforestation rates in the Amazon region by 70 percent over the next 10 years, following a call for international funding to prevent further loss of the Amazon rainforest. This plan was not considered in this study. If implemented, this plan could lead to biome distributions that are different than the ones projected here.

Scenarios

Deforestation

In the simulated deforestation scenarios, rainforest was converted to degraded grass (with deforested areas equal to 20 percent, 50 percent, and 100 percent of the original extent of the Amazon forest). The land cover change scenarios are from Sampaio et al. (2007) and Soares Filho et al. (2006), and consider that recent deforestation trends will continue; highways currently scheduled for paving will be paved; compliance with legislation requiring forest reserves on private land will remain low; and protected areas will not be enforced. Although extreme, it is important to evaluate scenarios of complete deforestation.

Climate Change

This study uses standard output, available through the Coupled Model Intercomparison Project phase 3 (CMIP3) multi-model dataset, from Coupled Ocean-Atmosphere GCMs for the IPCC AR4. (See details in table 5.1.)

Biome distribution was examined for the 21st century under emission scenarios A2 and B1,[3] which provide an outer envelope to the A1B scenario on the upper and lower bounds. Climate simulation for the end of the 20th century (20CM3) of each model is used to evaluate the models' anomalies. The precipitation and surface temperature monthly climatology for the Amazon (1961–1990) is obtained from work by Willmott and Matsuura (1998).

The climate change scenarios at regional scale (60-km resolution) were projected by the high-resolution MRI-JMA AGCM for the present time (1989–1999), near future (2015–2039) and future (2075–2099) for the IPCC SRES A1B emission scenario.

Models Used

The CPTEC Atmospheric Global Circulation Model

The CPTEC-INPE global atmospheric model (Cavalcanti et al. 2002) is used for the numerical simulations, at T062L42 spectral resolution (42 vertical levels, ~2° lat/lon horizontal resolution).[4] The land surface scheme is the SSiB (Xue et al. 1991). For each land grid point, a vegetation type (biome) is prescribed following the classification by Dorman and Sellers (1989) along with a set of physical, morphological, and physiological parameters.

Table 5.1. The Climate Models Referred To in This Analysis

Model	Institute (Country)	Resolution (Atmospheric component)
BCCR-BCM2.0	Bjerknes Centre for Climate Research (Norway)	T42L31 (approx. 2.8° lat/lon)
CCSM3	National Center for Atmospheric Research (USA)	T85L26 (approx. 1.4° lat/lon)
CGCM3.1(T47)	Canadian Centre for Climate Modelling and Analysis (Canada)	T47L31 (approx. 3.75° lat/lon)
CNRM-CM3	Météo-France/Centre National de Recherches Météorologiques (France)	T42L45 (approx. 2.8° lat/lon)
CSIRO-Mk3.0	CSIRO Atmospheric Research (Australia)	T63L18 (approx. 1.875° lat/lon)
ECHAM5/MPI-OM	Max Planck Institute for Meteorology (Germany)	T42L31 (approx. 2.8° lat/lon)
ECHO-G	Meteorological Institute of the University of Bonn (Germany), Institute of KMA (Rep. of Korea)	T30L19 (approx. 3.75° lat/lon)
GFDL-CM2.0	US Dept. of Commerce/NOAA/ Geophysical Fluid Dynamics Laboratory (USA)	2.° lat. x 2.5° lon., L24
GFDL-CM2.1	US Dept. of Commerce/NOAA/ Geophysical Fluid Dynamics Laboratory (USA)	2.° lat. x 2.5° lon., L24
GISS-ER	NASA/Goddard Institute for Space Studies (USA)	4° lat x 5° lon., L15
INM-CM3.0	Institute for Numerical Mathematics (Russia)	5° lat. x 4° lon, L21
IPSL-CM4.0	Institut Pierre Simon Laplace (France)	2.5° lat x 3.75° lon., L19
MIROC3.2(medres)	Center for Climate System Research (Univ. of Tokyo), National Institute For Environmental Studies, and Frontier Research Center For Global Change (Japan)	T42L20 (approx. 2.8° lat/lon)
MRI-CGCM2.3.2	Meteorological Research Institute (Japan)	T42L21 (approx. 2.8° lat/lon)
UKMO-HadCM3	Hadley Centre for Climate Prediction and Research/ Met Office (UK)	2.5° lat. x 3.75° lon. L19

Source: Table generated for the report by Sampaio et al. 2009.
Note: The models are selected from those in the Coupled Model Intercomparison Project phase 3 (CMIP3) multi-model dataset.

Potential Vegetation Model CPTEC-PVM2.0

A second vegetation model was also used to estimate the impacts of future climate and land cover change in the Amazon. The CPTEC-PVM2.0 (Lapola et al. 2009), a new version of the CPTEC (global) Potential Vegetation Model (Oyama and Nobre 2004), was applied to this end. It shows a particularly good performance over South America due to the consideration of seasonality as a determinant for the delimitation of forests and savannas. It also takes into account plants' physiological responses to seasonality (such as primary productivity) under varying atmospheric CO_2. The biome allocation relies mainly on the optimum net primary productivity (NPP) values for a given grid cell.

The determination of biome distribution through NPP is done based on numerous studies showing that different biomes have different average NPP (e.g., Sahagian and Hibbard 1998; Turner et al. 2006). However, in some cases NPP can be quite similar among biomes (such as for boreal forest and grassland) and in these cases variables other than NPP (e.g., coldest month temperature) are used for biome allocation. As a non-dynamic model, it calculates only equilibrium solutions based on long-term mean monthly climate variables. This is done concomitantly with a water balance submodel using climatologies of surface temperature and precipitation (1961–1990: Willmott and

Matsuura 1998), intercepted photosynthetically active radiation (IPAR) (1986–1995: Raschke et al. 2006), and atmospheric CO_2 (1961–1990: 350 ppmv) as inputs.[5]

CPTEC-PVM2.0 is forced by monthly precipitation, surface temperature and zonal wind inputs derived from ocean-atmosphere global climate models of the IPCC AR4, for the 1961–1990 period (actual climate) and three time slices in the 21st century (2010–2039, 2040–2069, and 2070–2099).

Simulations

Deforestation-Only Forcing

To evaluate the impact of a specific deforestation scenario, the climate and vegetation model results are used in sequence, according to the following steps. The climate model is run first, under the new deforestation scenario, and a climate condition is found that corresponds to this new surface configuration. Then, assuming that this resulting climate condition is sustained, the potential vegetation model is applied to find the new vegetation distribution in equilibrium with this climate. The resulting vegetation distribution will not necessarily reflect exactly the patterns in the deforestation scenario. That may happen because the new climate could support forest recovery or further savanna expansion, for example.

The CPTEC AGCM was integrated for 87 months, with five different initial conditions derived from five consecutive days of NCEP analyses, from October 14 to 18, 2002. Climate boundary conditions, including sea surface temperature, for experiments and control were used. The simulated deforestation was converted to degraded grass (for land cover change scenarios with deforested areas equal to 20 percent, 50 percent, and 100 percent of the original extent of the Amazon forest).

Climate-Change-Only Forcing

The CPTEC-PVM2.0 was used in three 20-year time slices of the 21st century: 2015–2034, 2060–2079, and 2085–2099 ("2025," "2075," and "2100" time slices, respectively), for the A2 and B1 scenarios of Greenhouse Gas Emissions (GHG) from 15 IPCC models.

Climate Change and Deforestation

The CPTEC-PVM2.0 was used by combining the methodology as described in "Deforestation-Only Forcing" and "Climate-Change-Only Forcing" above. A supposed deforestation of 20 percent is assumed in the "2025" time slice and a deforestation of 50 percent in the "2075" time slice. The climate anomalies from deforestation were combined with the anomalies of the IPCC scenarios, for each time slice. The total (100 percent) deforestation was not evaluated together with climate projections because of the major uncertainties associated with both extreme scenarios, and the lack of results from some of the climate models beyond the 21st century, when total deforestation would be assumed.

Climate Change, Deforestation, and Fire

The CPTEC-PVM2.0 was used as described above, adding the potential for occurrence of land use fires according to the method described in Cardoso et al. (2009) which is based on general relations between fire activity and climate factors, derived from analyses combining climate and soil hydrology variables to fires occurrence in the Brazilian Amazonia.

Biome Response to Different Forcings

Deforestation Only

The results from deforestation only simulations are summarized in figure 5.1. In the case of 20 percent deforestation (figure 5.1a), the biome-climate equilibrium state shows a reduction of forest area in Eastern and Southern Amazonia with savannas and tropical seasonal forest covering this region, and semi-desert area in Northeastern Brazil.

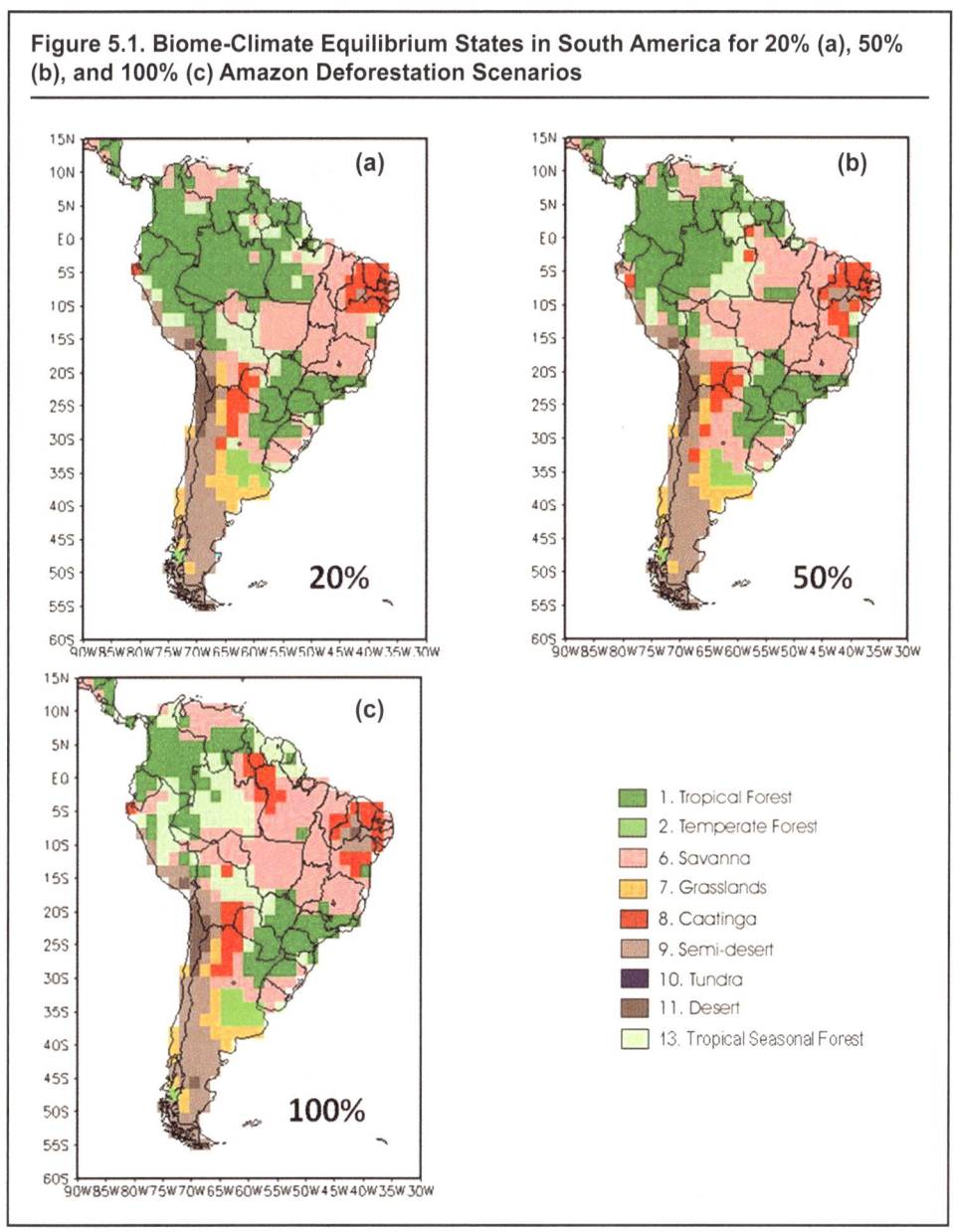

Figure 5.1. Biome-Climate Equilibrium States in South America for 20% (a), 50% (b), and 100% (c) Amazon Deforestation Scenarios

Legend:
- 1. Tropical Forest
- 2. Temperate Forest
- 6. Savanna
- 7. Grasslands
- 8. Caatinga
- 9. Semi-desert
- 10. Tundra
- 11. Desert
- 13. Tropical Seasonal Forest

Source: Figure generated for the report by Sampaio et al. 2009.

With 50 percent deforestation (figure 5.1b), the savannas and tropical seasonal forest areas cover a large part of Amazonia. The original forest cover is replaced by savannas and tropical seasonal forest and there is an expansion of the semi-desert area in Northeastern Brazil.

For the extreme case of 100 percent deforestation, the biome-climate equilibrium shows that most of Amazonia is covered by savannas and tropical seasonal forest (figure 5.1c). In Northeastern Amazonia there appears an area with dry shrubland (*caatinga*) and in Northeastern Brazil a large expansion of the semi-desert area. In general for the Amazon and Northeastern Brazil, there is replacement by drier climate biomes: savannas replacing forests, caatinga replacing savannas and semi-desert vegetation replacing dry shrubland. In these regions, the decrease in precipitation is more distinct in the dry season (June–October). In all cases, the average temperature near the surface increases with deforestation.

The results for specific regions are shown in figure 5.2. For Eastern Amazonia (figure 5.2a), the remaining area of tropical forest decreases with the expansion of the altered area at the initial condition. The remaining area of seasonal forest increases for 0–20 percent deforestation, decreases for 20–50 percent, and stabilizes for further deforestation. Savannas expand in all cases, but their rate of expansion is substantially higher, between 0 and 50 percent deforestation.

For Northwestern Amazonia (figure 5.2b), the decrease of the remaining area of tropical forest is higher after 50 percent deforestation. The expansion of the remaining area of seasonal forest exhibits a similar but inverse pattern, with pronounced expansion for deforestation fractions greater than 50 percent. The change in the remaining savanna area is less pronounced, and is greater for deforestation higher than 50 percent.

The patterns of change in Southern Amazonia (figure 5.2c) are similar to the patterns in the eastern part of the region, with tropical forests always decreasing with deforestation; savanna expansion is pronounced for percentages of altered area in the initial condition smaller than 50 percent. However, savannas show a slight decrease for higher values of altered initial areas.

For the entire region (figure 5.2d), the remaining area of tropical forest decreases almost linearly with deforestation. The other biomes expand in all cases; however, the expansion of savannas is more intense for 0–50 percent deforestation but stabilizes for higher values.

In Northeastern Brazil (figure 5.2e), there is a noticeable expansion of the semi-desert and a decrease of the remaining areas of savanna and caatinga. For Southern Brazil (figure 5.2f), the areas of all major biomes analyzed remain virtually unchanged.

Climate Change Only

The results of climate only impacts on biome equilibrium shifts are shown in figure 5.3. The results in figures 5.3 to 5.5 show grid points where more than 75 percent (at least 12) of the models coincide in projecting the future condition of biomes in relation to the current potential vegetation (75 percent consensus) for the different experiments. In these maps, "no consensus" means that fewer than 12 models agree with the transition. "Loss" means consensus for substitution of that biome class.

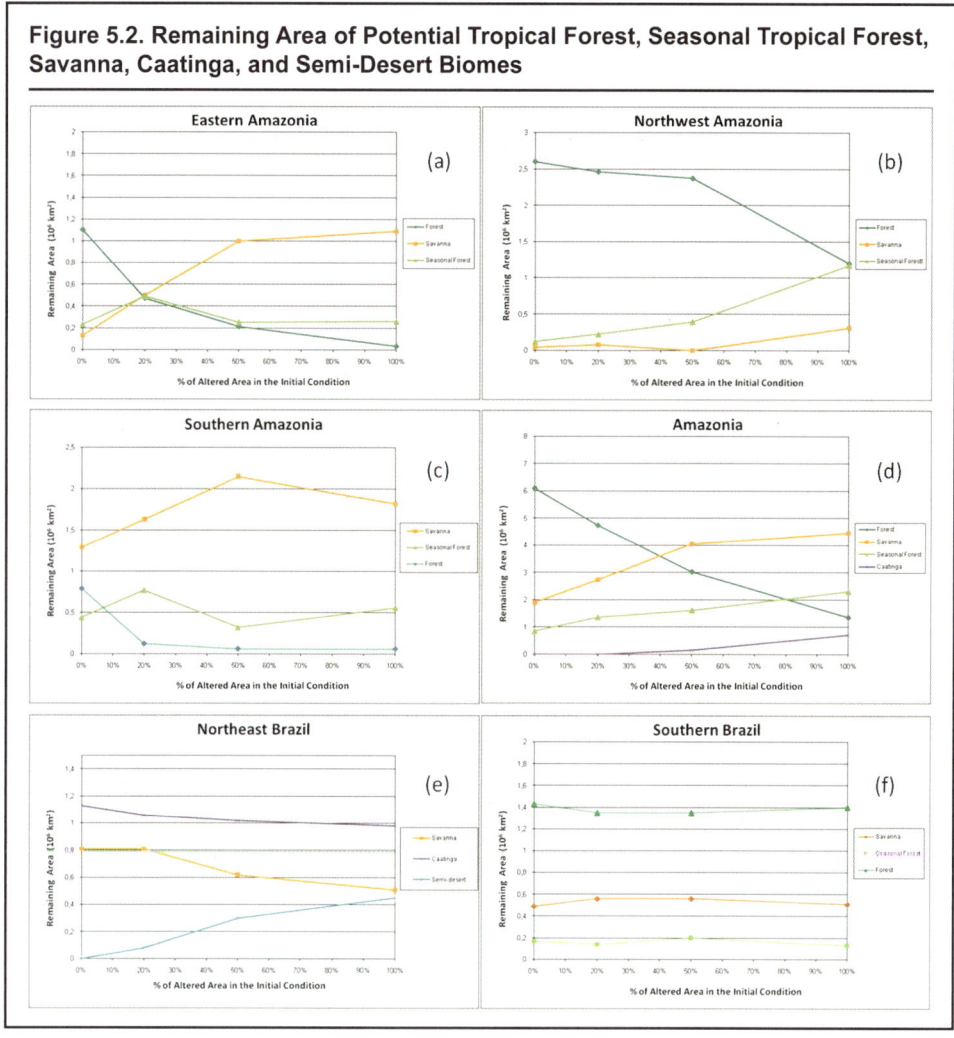

Figure 5.2. Remaining Area of Potential Tropical Forest, Seasonal Tropical Forest, Savanna, Caatinga, and Semi-Desert Biomes

Source: Figure generated for the report by Sampaio et al. 2009.
Note: Figure shows remaining area (10^6 km^2) of potential tropical forest, seasonal tropical forest, savanna, caatinga and semi-desert biomes for (a) Eastern Amazonia, (b) Northwestern Amazonia, (c) Southern Amazonia, (d) Amazonia as a whole, (e) Northeastern Brazil and (f) Southern Brazil, for land cover change scenarios with deforested areas equal to 20%, 50% and 100% of the original extent of the Amazon forest.

Climate change will potentially have important effects on the spatial patterns of the biomes' distribution in South America. The results for the Amazon under scenario A2 indicate consensus for change from tropical to seasonal forest by 2075, and from tropical to other biomes by 2100. In this region, for climate scenario B1, the results are less severe.

Climate Change and Deforestation

Combined global climate change and deforestation in Amazonia compound the effects on the spatial patterns of biome distribution in South America. The results of these in-

Figure 5.3. Grid Point for 75% Consensus on Future Condition of Tropical South American Biomes in Relation to Current Potential Vegetation

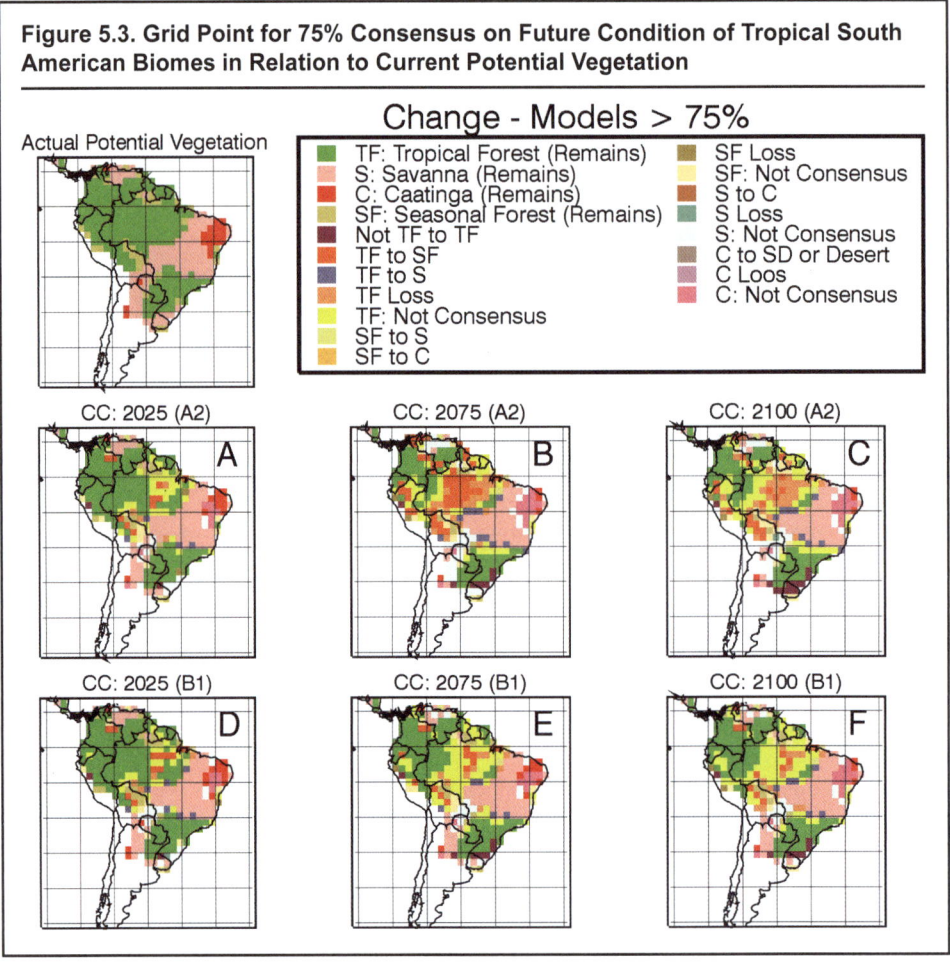

Source: Figure generated for the report by Sampaio et al. 2009.
Note: Maps show data for time slices (A) 2025, (B) 2075 and (B) 2100 for the A2 GHG emissions scenario, and (D), (E), (F) similarly for the B1 GHG emissions scenario. In these maps, "no consensus" means that fewer than 12 models agree with the transition. "Loss" means consensus for substitution of that biome class. SD means semi-desert.

teractions are shown in figure 5.4. Relative to the previous climate-change-only analysis, the results here generally show a larger area of tropical forest loss and noticeable savanna expansion for scenario B1 by 2075, under 50 percent deforestation. In Northeastern Brazil, for both climate scenarios, the area of consensus for changes from caatinga to other biomes is larger than in the previous results with expansion of caatinga to the North and Northwest. In Southern Brazil, the results are similar to the climate-change-only analysis and project consensus for expansion of tropical forest over areas of potential savannas.

Climate Change, Deforestation, and Fire

Figure 5.5 displays results considering the combination of global climate change, deforestation, and fires. An important feature in these results is that the effect of including the potential for fire occurrence is greater in the period before 2025 for both climate scenarios. Fire potential corroborates the consensus for transitions from tropical forest

Figure 5.4. Grid Point for 75% Consensus on Projecting the Future Condition of Tropical South American Biomes in Relation to Current Potential Vegetation

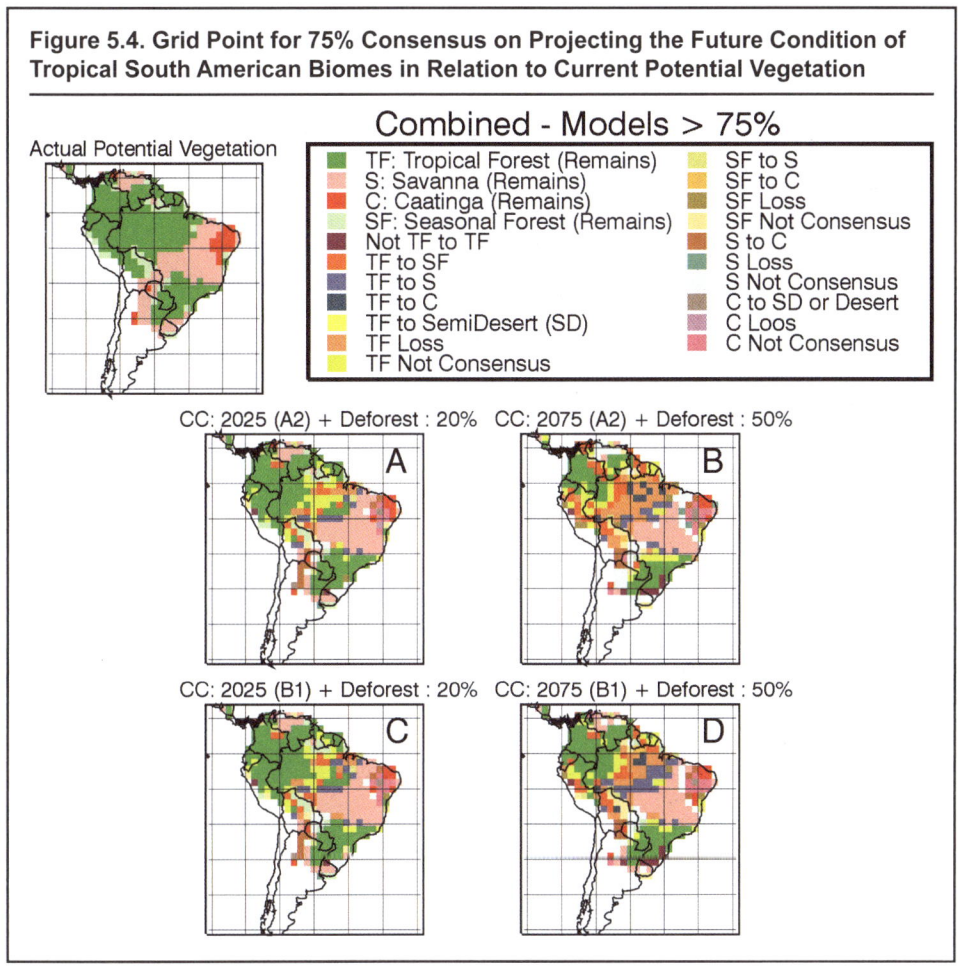

Source: Figure generated for the report by Sampaio et al. 2009.
Note: Maps show data for time slices (A) 2025 + 20% deforestation and (B) 2075 + 50% deforestation for the A2 GHG emissions scenario, and (C), (D) similarly for the B1 GHG emissions scenario. In these maps, "no consensus" means that fewer than 12 models agree with the transition. "Loss" means consensus for substitution of that biome class.

to savanna in the South and Southeast region of Amazonia. For Amazonia as a whole, the remaining tropical forest area for time slice 2025 progressively reduces as climate change impacts, deforestation, and fire are combined relative to its original extension. The projected remaining Amazon rainforest biome by 2100, under scenario A2, is about one third of the original.

Major impacts are projected in Eastern Amazonia. The combined effects of climate and deforestation result in a severe decrease of the rainforest biome, in relation to its original extension for forest area. The remaining forest biome, by 2075, accounting for 50 percent deforestation and or the effects of fires, is about 5 percent. This is the largest relative decrease in the entire basin.

The Northwest projections indicate the smallest relative decrease of tropical forest biome for the entire basin. The potential for fire occurrence was found to be low in this

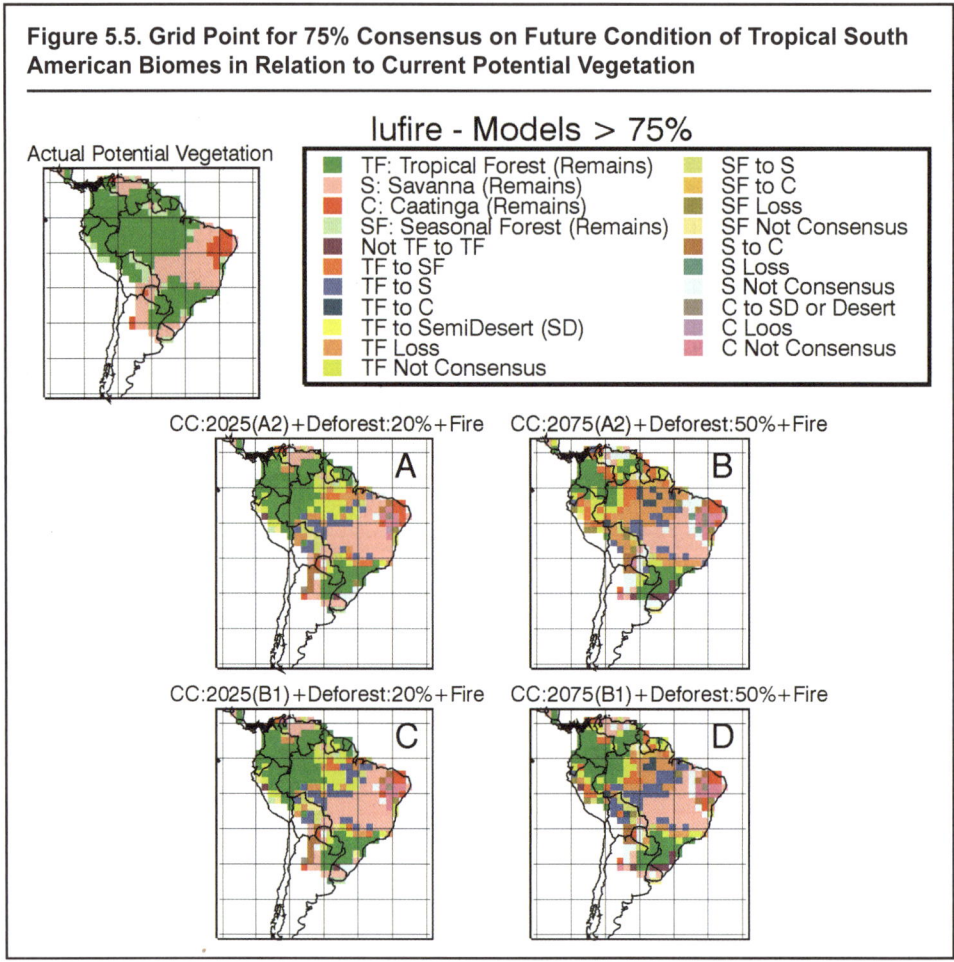

Figure 5.5. Grid Point for 75% Consensus on Future Condition of Tropical South American Biomes in Relation to Current Potential Vegetation

Source: Figure generated for the report by Sampaio et al. 2009.
Note: Maps show data for time slices (A) A2 GHG emissions scenario 2025 + 20% deforestation + fire; (B) A2 GHG emissions scenario 2075 + 50% deforestation + fire; (C) B1 GHG emissions scenario 2025 +20% deforestation + fire; and (D) B1 GHG emissions scenario 2075 + 50% deforestation + fire. In these maps, "no consensus" means that fewer than 12 models agree with the transition. "Loss" means consensus for substitution of that biome class.

sub-region. The decrease of the tropical forest area is smaller than 10 percent for time slice 2025. However, even for this region, under scenario A2 there would be a significant decrease of tropical forest area by 2100, which most of the change would be achieved by 2075 (about 50 percent of the remaining biome).

For Southern Amazonia, the analysis indicates a relative increase in the area of savanna. Deforestation contributes to this configuration with a drier and hotter climate, favoring savanna expansion in replacement of tropical forests. In this region, fire has an important effect. For scenario A2 (and B1), and time slices 2025 and 2075, the projection indicates a net increase in the area of savanna, ranging from 30 percent to 87 percent.

For the Northeast the analysis indicates a slight relative increase (7 percent) in caatinga, resulting from hotter and drier climate.

Table 5.2. Consensus Results for Remaining Share of Reference Biome Compared to Baseline (=100) by Geographical Domain, from 2025 to 2100 under Scenario A2

Domain	Reference biome	Climate change impact by 2025	+20% defores-tation	+ fire impact	Climate change impact by 2075	+50% defores-tation	+ fire impact	Climate change impact by 2100
EA	Rainforest	56(29)	29(11)	25(10)	17(22)	2(2)	2(2)	26(26)
NWA	Rainforest	94 (6)	90(6)	90(6)	52(20)	48(18)	48(18)	60(22)
SAz	Savanna	130(18)	138(21)	187(7)	136(18)	143(25)	164(23)	134(20)
NEB	Caatinga	94(18)	93(9)	93(9)	92(34)	107(34)	107(34)	82(33)
SB	Rainforest	110(±5)	108(±6)	106(±6)	123(±5)	120(±5)	116(±7)	129(±5)

Source: Table generated for the report.
Note: Similar but smaller transitions are found for scenario B1. Parentheses display the standard deviation of remaining share of reference biome.

In Southern Brazil the combination of factors leads to a decrease in the forest area, with a strong effect derived from forest fires. In general, fire potential strengthens the consensus for transitions from tropical forest to savanna in the Southern and Southeastern regions of Amazonia (table 5.2).

Summary of Regional Results

Table 5.3 summarizes the results by region under analysis (geographical domains, as defined in chapter 1) including the net changes in vegetation carbon and the risk of Amazon dieback induced by climate forcing under A1B scenario. The summary portrays the different momentums of land use change driven by the changes in vegetation carbon and anticipated shifts in biome driven by changes in bioclimatic conditions.

From the results of the analysis, the combined effects of climate change, deforestation and fire events will have very serious impacts on the rainforest biome throughout

Table 5.3. Regional Impacts

Geographic domain (defined in figure 1.1)	Current vegetation carbon density (kg C m^{-2})	Risk of dieback (%) (probability of a 25% reduction in veg. carbon)	Change in vegetation carbon by two best-ranked models (kg C m^{-2}) (*)	Potential biome shift caused by climate change with deforestation at 50% and occurrence of fires by 2075 [CPTEC-PVM]
EA— Eastern Amazonia	12	8	-2 to -6	Savanna from 20% to 40% of area; Significant reduction in Tropical Rainforest.
NWA— North Western Amazonia	15	3	-5 to -6	No significant change in cover of Tropical Rainforest or Seasonal Forest.
SAz— Southern Amazonia	10	62	-4 to -5	No major change in cover of Savanna; further reduction in Tropical Rainforest biome from 10% to less than 5%; slight increases in Seasonal Forest.
NEB— Northeastern Brazil	3	19	-1 to -2	Reduction of Seasonal Forest from 5% to less than 1%, and transitions of Savanna to Caatinga and Caatinga to desert.
SB— Southern Brazil	10	2	-1 to -3	Reduction in cover of grasslands and corresponding increase in Savanna and Rainforest.

Source: Table generated for the report.

the Amazon basin. While the impacts from climate change result from global contributions, the deforestation and induced fires are aspects that can be addressed within the country.

In fact, the results support the view that there is a need to arrest the deforestation of the Amazon basin as it reinforces the stress induced by the anticipated impacts from climate change. However, as table 5.3 shows, implications are somewhat different depending on the geographical domain under analysis.

Notes

1. Oyama and Nobre (2003) showed that it is possible to have two biome-climate equilibrium states in tropical South America. One equilibrium state corresponds to the current vegetation distribution where the tropical forest covers most the Amazon Basin and the second corresponds to stable biome-climate equilibrium with savannas covering eastern Amazonia and semi-deserts in Northeastern Brazil.

2. There is no model that combines the dynamic and coupled features of LPJmL and CPTEC-CPVM and thus these two tools are used, where they can best be applied.

3. These represent a plausible range of conditions over the next century. In scenario B1, the atmospheric CO_2 concentration in year 2100 (2025, 2075) reaches a level of 550 ppm (410 ppm, 480 ppm); in A2 the corresponding value is 730 ppm (410 ppm, 520 ppm) (IPCC, 2007).

4. The CPTEC-INPE AGCM (atmospheric global circulation model) available at INPE was developed at the Brazilian Center for Weather Forecast and Climate Studies (CPTEC) based on the CPTEC/COLA (Center for Ocean-Land-Atmosphere Studies) GCM described by Cavalcanti et al. (2002). It has been shown that the model simulates reasonably well the main features of global climate, as well as the seasonal variability of the main atmospheric variables.

5. Water balance routine is nearly the same as in CPTEC-PVM (Oyama and Nobre 2004) based on Willmott et al. (1985), although canopy resistance r_c (1/canopy conductance g_c) is calculated in terms of NPP and atmospheric (CO_2), based on the formulation by Collatz et al. (1991), which is used by several DGVMs (Sitch et al. 2008) and GCM surface schemes (e.g., Sellers et al. 1996)). The canopy resistance is used to calculate evapotranspiration (hereafter E) according to Penman-Monteith's equation. This formulation enables a two-way interaction of water cycle and plant physiology. CPTEC-PVM2 also considers a simple parameterization of lightning-induced fires in savannas based on the study by Cardoso et al. (2008). Fire occurrence is regarded as dependent on the availability of natural ignition source (using 850 hPa zonal wind as a proxy to lightning) and fuel moisture (through soil water level). The fertilization effect in this model is considered 25 percent of the total CO_2 atmospheric concentration for each emission scenario (A2 and B1). Global and regional NPP simulated by CPTEC-PVM2 are quite comparable to that from observations and also from other NPP models. Biome distribution is evaluated against an analysis of natural vegetation (Lapola et al. 2008) and results in a global kappa statistic (Monserud and Leemans 1992) of 0.53, and an agreement fraction of 57 percent (Lapola et al. 2009).

CHAPTER 6

Conclusions

The effects of climate change can modify the functioning and structure of the Amazon basin. With rising atmospheric CO_2 concentrations, climate change will lead to a substantial warming in Amazonia during the current century, reaching levels that are highly likely to affect the remaining forests throughout the region.

It is expected that the Amazon region will experience an intensification of the water cycle with increased occurrence of heavy rainfall and consequent flooding and a lengthening of drought periods. Specific projections for indexes reflecting increased rainfall and dry period extremes, prepared through the application of the high-resolution MRI AGCM with the use of the Earth Simulator for the Amazon basin, are included in the report. Likewise, the region's hydrology will be affected, with increased stream flows during wet periods and lower than current stream flows for dry periods in major rivers of the region.

However, there is considerable uncertainty over future rainfall. Most climate models project substantial changes in rainfall patterns, but these do not coincide: some models project increases, others project decreases, and the spatial pattern of these changes also varies between the models.

To estimate the risk of Amazon drying, probability density functions for future rainfall were derived using two procedures described in the report. These indicate a strong likelihood of Northwestern Amazonia becoming wetter in the December to May period, and an increased probability of 2005-like drought conditions in Southern Amazonia (from 1 event in 100 years currently, to 1 in 17 years) toward the end of the century. In other regions the probability of change is less significant due to continuing discrepancies among climate model simulations in this region.

The climate model PDFs based on rainfall were combined with forest simulations to produce PDFs of changes in vegetation carbon in the 21st century. In the absence of CO_2 fertilization, a reduction in vegetation carbon is anticipated in most of the geographical domains. The analysis concludes that there is a substantial probability of Amazon dieback.

When leaf physiology-based CO_2 fertilization default factors are used, the likelihood of this reduction is substantially lower, and the probability of increases in vegetation carbon is non-negligible. However, in tropical, mature forest ecosystems, under pronounced nutrient constraints typical of poor soil conditions in the Amazon basin, there is great uncertainty that CO_2 fertilization may play such an effective role. Thus, in the absence of solid information, such as ecosystem-wide CO_2 fertilization experiments, the assumption that CO_2 fertilization will be an important factor positively-affecting ecosystem resilience of the Amazon cannot be used presently as a basis for sound policy advice. Reducing this uncertainty, using a combination of estimates of the Amazonian carbon sink and trends in river runoff, is therefore a priority for a follow-on study.

The impacts of deforestation, climate change, and fires were estimated using the outputs of 15 GCMs and the CPTEC-PVM, using a consensus approach, where only changes in biome with 75 percent or higher agreement are identified. When the effects of climate change, deforestation, and fire are combined, using a potential vegetation model (CPTEC-PVM) specifically developed and applied in the South American context, major shifts in biomes are predicted.

For Amazonia as a whole, the remaining tropical forest area relative to its original extension is progressively reduced as climate change impacts, deforestation and fire effects are combined. Substantial impacts are already projected by 2025 and the situation worsens by 2050. The effect of climate change alone would contribute to reduce the extent of the rainforest biome by one third by the end of the century.

Major impacts are projected in Eastern Amazonia. The combined effects of climate and deforestation result in a severe decrease of the rainforest biome, in relation to its original extension of forest area. The remaining forest biome, by 2075, accounting for 50 percent deforestation and/or the effects of fires, is about 2 percent. This is the largest relative decrease in the entire basin.

The Northwest projections indicate the smallest relative decrease of tropical forest biome for the entire basin. The potential for fire occurrence was found to be low in this sub-region. The decrease of the tropical forest area is smaller than 10 percent for time slice 2025. However, even for this region, under scenario A2 there would be a significant decrease by 2100 (about 60 percent of the remaining biome).

For Southern Amazonia, the analysis indicates a relative increase in the area of savanna. The deforestation contributes to this configuration with a drier and hotter climate, favoring savanna expansion in replacement of tropical forests. In this region, fire has an important effect. For scenario A2 (and B1) and time slices 2025 and 2075, the projection indicates a net increase in the area of savanna, ranging from 30 percent to 87 percent.

For the Northeast the analysis indicates a slight relative increase (7 percent) in caatinga over savanna, resulting from hotter and drier climate.

In the South and Southeast Amazonia the combination of factors leads to a decrease in the forest area, with a strong effect derived from forest fires. In general, fire potential strengthens the consensus for transitions from tropical forest to savanna in Southern and Southeastern Amazonia.

Biome projections for the end of the century in tropical South America show a variety of results, depending not only on the climate scenario, but also on the effect and the level of CO_2 fertilization on photosynthesis. In summary, the results for numerical experiments indicate potential for:

- Transition from tropical to seasonal forest and savanna in the East/Southeast of Amazonia, related mainly to a projected decrease in precipitation in the dry season and increase in average annual temperature.
- In the Northwest of Amazonia, the results indicate maintenance of tropical forest, related to projected increase in precipitation and evapotranspiration.
- Different than the global models, the regional climate model for the Southern Amazon projects a smaller decrease in precipitation, which in turn does not support transitions to sparse biomes such as caatinga.

- In Northeast Brazil the regional climate model also projects a slight increase in the precipitation at the beginning of the rainy season, which translates into a slight substitution of areas of caatinga by savanna.
- In the Southern Brazil and the La Plata Basin, the climate projections indicate increase in precipitation which favors the transitions to biomes with denser vegetation, such as transitions from grassland to tropical forest, or caatinga to savanna.

The synergistic combination of regional climate impacts due to deforestation and climate change resulting from global warming, and the potential for higher fire occurrence adds considerably to the vulnerability of tropical forest ecosystems in the study region. In several cases, the remaining biome is savanna which is generally more adapted to hotter climates with marked seasonality in rainfall (long dry seasons), where fire naturally plays an important ecological role.

Losses linked to Amazon dieback would show up in agriculture, forestry, and power generation, among other sectors of economic activity. However, a full accounting of losses would need to include those incurred in environmental services (fresh water, oxygen, biodiversity, ecosystem integrity, services to other species) or the value of the lost genetic information through a major collapse of the system. The loss of the Amazon rainforest and its associated ecosystems has an intrinsic value that is not amenable to quantification but certainly exceeds any accounting made with current economic tools. Nevertheless, a better valuation of the financial and natural capital represented by the Amazon ecosystem is required as well as a more comprehensive assessment of the economic implications of its potential dieback; this should also be part of a follow-up.

All of these results indicate the need to avoid reaching a point of GHG emissions that would result in an induced Amazon loss. The current emissions trajectory may result in a high risk of incurring these losses during this century. Thus, Amazon dieback should be considered a threshold for dangerous climate change. Likewise, the estimated combined effects of climate impacts and deforestation on the integrity of the Amazon strongly suggest that deforestation should be rapidly reduced.

We emphasize that biome projections performed in this study are based on the land-use patterns we currently observe in the study region. In fact, Brazil has announced a plan to reduce deforestation rates in the Amazon region by 70 percent over the next ten years, which was not considered in this study. If this plan would be implemented, it could lead to biome distributions that are different than the ones projected here.

In sum, climate impacts have the potential to disrupt the functioning and structure of the Amazon forest beyond the natural capacity of regeneration or recovery.

Next Steps

The conclusions reached in the report have significant implications. The risk of Amazon dieback, in the absence of any significant CO_2 fertilization effect at ecosystem level, has been found to be substantial for key regions of the Amazon basin. The potential shift of an equilibrium state of the basin toward biomes with less biomass should be of great concern. This potential shift is likely to be exacerbated by the combined effects of deforestation, climate change and associated increases in the likelihood of fires.

It is recommended that additional efforts be made to ascertain the potential role, if any, of CO_2 fertilization and its effect on growth, carbon stocking, and resilience at ecosystem level for the Amazon basin. Direct CO_2 effects at ecosystem level are the key unknown in assessing the risk of Amazon forest dieback under 21st century climate change. Reducing this uncertainty, using a combination of estimates of the Amazonian carbon sink (Phillips et al. 2009) and trends in river runoff (Gedney et al. 2006), is therefore a priority for a follow-on study.

Furthermore, the findings indicate that differences between the total impacts of climate change caused by CO_2 and other climate forcing agents (e.g., increases in methane or reductions in sulphate aerosols) have implications for international climate policy, which currently treats all radiative forcings as equally damaging. In contrast, this study shows clearly that the risk of Amazon forest dieback is many times greater if climate change arises from agents other than CO_2; this differential effect should also be further assessed.

Finally, the quantification of the potential economic consequences of Amazon dieback require additional efforts to monetize all the implications derived from major changes in the global and regional environmental and economic services provided by the Amazon basin. A more comprehensive evaluation would certainly substantiate the justification of any remedial actions.

Appendixes

Appendix A. IPCC—Emissions Scenarios

In 1992, for the first time the IPCC released emission scenarios for use in driving global circulation models to develop climate change scenarios.

In 1996, the IPCC decided to develop a new set of emissions scenarios (the Special Report on Emissions Scenarios, or SRES), which provided input to the IPCC's Third Assessment Report (TAR) in 2001. The SRES scenarios were also used for the Fourth Assessment Report (AR4) in 2007. Since then, the SRES scenarios have been subject to discussion because emissions growth since 2000 may have made these scenarios obsolete. It is clear that the IPCC's Fifth Assessment Report will develop a new set of emissions scenarios.

The Amazon dieback task used several of the IPCC's SRES emissions scenarios for its analyses (i.e., A1B, B2, and B1). The following paragraphs provide a brief background on the IPCC SRES scenarios and show the expected range of temperature increase toward the end of the 21st century under each of these scenarios.

The SRES scenarios cover a wide range of the main driving forces of future emissions, from demographic to technological and economic developments. None of the scenarios includes any future policies that *explicitly* address climate change, although all scenarios necessarily encompass various policies of other types and for other sectors. The set of SRES emissions scenarios is based on an extensive literature assessment, six alternative modeling approaches, and an "open process" that solicited wide participation and feedback from many scientific groups and individuals. The SRES scenarios include a range of emissions of all relevant greenhouse gases (GHGs) and sulfur and their underlying driving forces.

As an underlying feature of all emissions scenarios, the IPCC developed four different narrative storylines to describe the relationships between emission-driving forces and their evolution over time. Each storyline represents different demographic, social, economic, technological, and environmental developments. Each emissions scenario represents a specific quantitative interpretation of one of the four storylines. All the scenarios based on the same storyline constitute a scenario "family."[1]

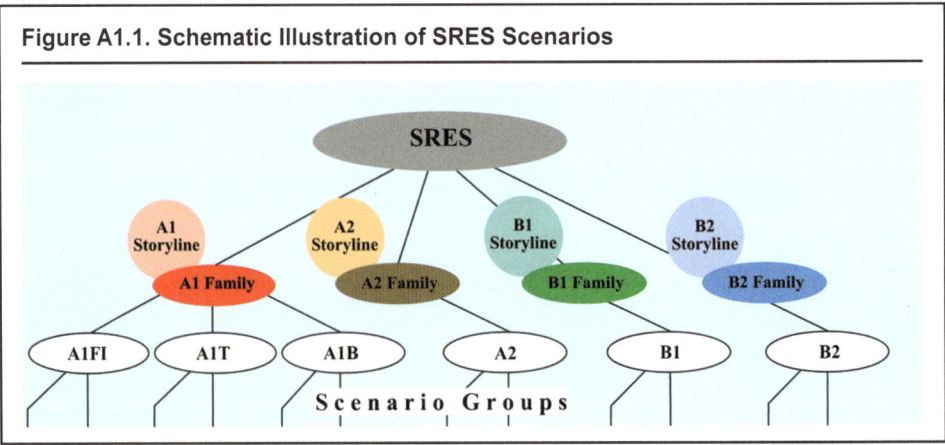

Figure A1.1. Schematic Illustration of SRES Scenarios

Source: IPCC 2000, modified.

The **A1 storyline** and scenario family describes a future world of rapid economic growth, global population peaks by mid-21st century and declines thereafter, and the rapid introduction of new and more efficient technologies. Major underlying themes of the A1 storyline are convergence among regions, capacity building, and increased cultural and social interactions, with a substantial reduction in regional differences in per capita income. The A1 scenario family develops into three groups that describe alternative directions of technological change in the energy system. The three A1 groups are distinguished by their technological emphasis: **fossil intensive (A1FI)**, **non-fossil energy sources (A1T), or a balance across all sources (A1B)**.

The **A2 storyline** and scenario family describes a rather heterogeneous world. The underlying theme is self-reliance and preservation of local identities. Global population increases continuously. For the most part, economic development is regionally oriented, and per capita economic growth and technological change are more fragmented and slower than in other storylines.

The **B1 storyline** and scenario family describes a convergent world with the same global population that peaks in mid-century and declines thereafter, as in the A1 storyline, but with rapid changes in economic structures toward a service and information economy, with reductions in material intensity, and the introduction of clean and resource-efficient technologies. The emphasis is on global solutions to economic, social, and environmental sustainability, including improved equity, but without additional climate initiatives.

Table A1.1. Projected Global Average Surface Warming and Sea Level Rise at the End of the 21st Century According to Different SRES Scenarios

Case	Temperature change (degrees centigrade at 2090-2099 relative to 1980-1999) [a,d]	Sea level rise (meters at 2090-2099 relative to 1980-1999	Model-based range (excluding future rapid dynamical changes in ice flow)
	Best estimate	*Likely* range	
Constant year 2000 concentrations [b]	0.6	0.3-0.9	Not available
B1 scenario	1.8	1.1-2.9	0.18-0.38
A1T scenario	2.4	1.4-3.8	0.20-0.45
B2 scenario	2.4	1.4-3.8	0.20-0.43
A1B scenario	2.8	1.7-4.4	0.21-0.48
A2 scenario	3.4	2.0-5.4	0.23-0.51
A1FI scenario	4.0	2.4-6.4	0.26-0.59

Source: IPCC 2007.

Notes: a) Temperatures are assessed best estimates and likely uncertainty ranges from a hierarchy of models of varying complexity as well as observational constraints.

b) Year 2000 constant composition is derived from Atmosphere-Ocean General Circulation Models (AOGCMs) only.

c) All scenarios above are six SRES marker scenarios. Approximate CO2-eq concentrations corresponding to the computed radiative forcing due to anthropogenic GHGs and aerosols in 2100 (see p. 823 of the Working Group I TAR) for the SRES B1, AIT, B2, A 1 B, A2, and A1FI illustrative marker scenarios are about 600, 700, 800, 850, 1250, and 1550ppm, respectively.

d) Temperature changes are expressed as the difference from the period 1980-1999. To express the change relative to the period 1850-1899 add 0.5°C.

The **B2 storyline** and scenario family describes a world in which the emphasis is on local solutions to economic, social, and environmental sustainability. It is a world with global population continuously increasing at a rate lower than that of A2, intermediate levels of economic development, and less rapid and more diverse technological change than in the B1 and A1 storylines. While the scenario is also oriented toward environmental protection and social equity, it focuses on local and regional levels.

Table A1.1 summarizes the likely temperature changes under each of the scenarios described above. It shows that B2 would lead to a temperature change of approximately 2.4° C toward the end of the century, under A1B the temperature change is estimated to be 2.8° C, while A2 is more extreme with a 3.4° C projected change.

It is important to note that the projected surface temperature changes toward the end of the 21st century exhibit a broad range of likely estimates, as shown by the bars next to the right panel of figure A1.2.

Figure A1.2. Scenarios for GHG Emissions from 2000 to 2100 (in the Absence of Additional Climate Policies) and Projections of Surface Temperatures

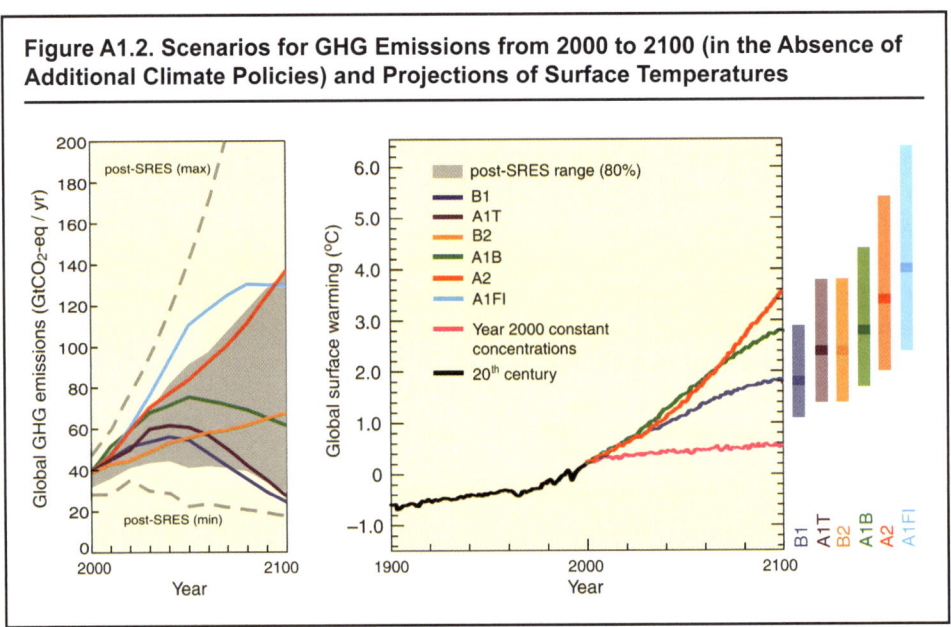

Source: IPCC 2007.

Notes: **Left Panel:** Global GHG emissions (in GtCO$_2$-eq) in the absence of climate policies: six illustrative SRES marker scenarios (colored lines) and the 80th percentile range of recent scenarios published since SRES (post-SRES) (gray-shaded area). Dashed lines show the full range of post-SRES scenarios. The emissions include CO$_2$, CH$_4$, N$_2$O and F gases.

Right Panel: Solid lines are multi-model global averages of surface warming for scenarios A2, A1B and B1, shown as continuations of the 20th-century simulations. These projections also take into account emissions of short-lived GHGs and aerosols. The pink line is not a scenario, but is for Atmosphere-Ocean General Circulation Model (AOGCM) simulations where atmospheric concentrations are held constant at year 2000 values. The bars at the right of the figure indicate the best estimate (solid line within each bar) and the likely range assessed for the six SRES marker scenarios at 2090–2099. All temperatures are relative to the 1980–1999 period.

Note

1. For each storyline, several different scenarios were developed using different modeling approaches to examine the range of outcomes arising from a range of models that use similar assumptions about driving forces.

Appendix B. Development of Probability Density Functions (PDFs) for Future Amazonian Rainfall

The study estimates Probability Density Functions (PDFs) for future rainfall in five regions of Amazonia by weighting the predictions of the 24 Coupled Model Inter-comparison Project (CMIP3) General Circulation Models (GCMs). Presented are two separate sets of model weightings. In the first set of model weightings, the models are rated according to their relative abilities to reproduce the mean and variability of the observed rainfall in each season. In the second set of model weightings, the models are rated according to their relative abilities to reproduce the mean and variability of two sea surface temperature indexes: the Atlantic North-South Gradient (ANSG) and the Pacific East-West Gradient (PEWG). In both cases, the relative weighting of the climate models is updated sequentially according to Bayes' theorem, based on the biases in the mean of the predicted time series and the distributional fit of the bias-corrected time series as measured by the Kolmogorov-Smirnov statistic, D. Depending on the assessment criteria, the season and the region, very different rankings of the GCMs are found, with no individual model doing well in all cases. In many cases there is also no significant correlation between the quality of a model's 20th century rainfall simulation and its prediction of rainfall change in the 21st century.

However, in some regions and seasons there are significant shifts in the derived rainfall PDFs, with the Southern Amazonia and Eastern Amazonia regions predicted to get wetter in the December–February period of the wet season, and an increased probability of 2005-like drought conditions in Northwest Amazonia in the June–August period. Using a combination of the relative model weightings for each season, the study also derives a set of overall model weightings for each region, which can be used to produce PDFs of forest biomass from the simulations of the LPJ model.

Weighting of 24 CMIP3 General Circulation Models (GCMs)

The various model projections were weighted based on the ability of each model to produce key aspects of the observed climate. In this way it is hypothesized to reach more robust predictions by emphasizing the results from more realistic models and de-emphasizing the results produced by less realistic models. The approach described is to construct a probabilistic prediction based on a weighted sum of the predictions of individual GCMs, using a Bayesian approach. The weight assigned to each GCM will be referred to as the probability of the model and will generate a PDF over the set of models. Bayes' theorem allows the model probabilities to be modified each time we consider the ability of the models to simulate some relevant aspect of current climate (such as rainfall in each season). This updating of the PDF is achieved by comparing time series of past observations with time series of model simulations for each variable. In this study the models are weighted based on their ability to simulate both the mean state and the variability (i.e., the statistical distribution) of current climate. In other words, the aim is to down-weight those models that simulate a climate whose mean value is far from the observed mean, or a climate whose statistical distribution is a poor fit to the observed distribution, even when any bias in the mean value has been corrected.

The procedure can be summarized as follows (mathematical details can be found in Jupp & Cox 2008 and in Cox & Jupp 2009):

   i.   assign equal probability to all models: a uniform prior PDF;

   ii.   choose a climatic variable of interest and update the model PDF based on the fit between model simulations and observations for this variable;

   iii. treat the current model PDF as a new prior, and repeat steps (i) and (iii) as required;

   iv. obtain a final posterior PDF for the models.

This procedure is used to estimate PDFs for future rainfall in each of the five regions of Amazonia, using rainfall simulations produced by the 24 GCMs available in the archive of the CMIP3 (see table A2.1).

Data

Two types of data are used in this study: observational data for the 20th century alone and model-based data for the 20th and 21st centuries. The data taken to represent the "true" state of the climate are taken from the CRU TS 3.0 archive. These data are available at http://www.cru.uea.ac.uk/cru/data/hrg-interim/. GCM data are taken from the CMIP3 multi-model archive, in the Climate of the 20th Century experiment. There are 24 models, as listed in table A2.1. These data are available at https://esg.llnl.gov:8443/.

    Temporal coverage: For assessment of 20th century climate, data are considered at a monthly resolution for the January 1901–December 1999 period (this is the longest period for which data are available from all sources.) Similarly, model predictions for the 21st century are considered at a monthly resolution for the January 2001–December 2099 period.

    Spatial coverage: Time series are created by taking spatial averages of monthly data in a total of nine spatial windows (table A2.2). Five of these windows are in South America and four cover the oceans. Sea surface temperatures in the four oceanic windows are used to define the ANSG and the PEWG.

Results

Relative model weightings are derived for each season (DJF, MAM, JJA, SON) as well as for the entire calendar year ("all") and for each of the five selected regions of Amazonia as listed in table A2.2.

    In each case, the weighting makes use of both the bias in the mean rainfall and the Kolmogorov-Smirnov statistic of the bias-corrected rainfall (through index D) to down-weight the models. The figures below show the relative model weightings derived for each region and for each of these three-month periods. These figures also show the overall model weightings that result from combining the weightings from each of the four three-month periods. These overall weightings are to be used to produce PDFs from the forest projections produced by the LPJ model.

    It is clear that the relative ranking of GCMs varies significantly with region and season. In any one region it is also unusual for a given model to simulate rainfall accurately in all four seasons. As a result, models that simulate each season well tend to dominate the overall weighting. The relative weightings by season tend to be much more evenly distributed among the models, except for the April–June period in Northeast Brazil, which is especially badly simulated by all models except 6 and 19 (as a result, these models dominate the overall weighting in this region).

    The relative model weights can be used to derive rainfall PDFs for each region and season.

The PDFs and Cumulative Distribution Functions (CDFs) are shown in figures A2.9 to A2.13, in each case for the 2001–2031 (black) and 2069–2099 (red) periods. Also shown is the difference between the prior distributions (dotted lines), which were derived by assuming all models to be equally likely, and the posterior distributions (continuous lines), which were derived using our Bayesian procedure. As hoped, in most cases weighting the models by their relative abilities to produce the current rainfall leads to sharper PDFs. This is most obviously demonstrated in Eastern Amazonia in December to February (see figure A2.9, top right panel). In some cases, the Bayesian weighting also shifts the most likely future rainfall significantly, for example in the October–December period in Northwest Amazonia (see figure A2.11, lower right panel).

Despite the gains associated with weighting models in this way, the overall uncertainties in future Amazonian rainfall remain significant. The few regions and seasons where significant changes in the rainfall PDF are predicted to occur during the 21st century include: Eastern Amazonia and Southern Amazonia, which are predicted to get wetter in the wet season (see top-right panels of figures A2.9 and A2.12), and Northwest Amazonia, which has an increased probability of 2005-like drought conditions in the July–September pre-wet season (see bottom-left panel of figure A2.11).

Table A2.1. The Climate Models Referred To in This Study

Model Identifier	Model Name	Model Identifier	Model Name
a	bccr_bcm_2_0	m	ingv_echam4
b	cccma_cgcm3_1	n	inmcm3_0
c	cccma_cgcm3_1_t63	o	ipsl_cm4
d	cnrm_cm3	p	miroc3_2_hires
e	csiro_mk3_0	q	miroc3_2_medres
f	csiro_mk3_5	r	miub_echo_g
g	gfdl_cm2_0	s	mpi_echam5
h	gfdl_cm2_1	t	mri_cgcm2_3_2a
i	giss_aom	u	ncar_ccsm3_0
j	giss_model_e_h	v	ncar_pcm1
k	giss_model_e_r	w	ukmo_hadcm3
l	iap_fgoalsl_0_g	x	ukmo_hadgem1

Source: Table generated for the report by Cox and Jupp 2009.
Note: The models are those in the World Climate Research Programme's (WCRP's) Coupled Model Intercomparison Project phase 3 (CMIP3) multi-model dataset in the "Climate of the 20th Century" experiment https://esg.llnl.gov:8443/.

Table A2.2. The Spatial Windows Relevant To This Study

Region	Identifier	Longitude	Latitude
Eastern Amazonia	EA	55°W - 45°W	5°S - 2.5°N
Northwest Amazonia	NWA	72.5°W - 60°W	5°S - 5°N
Northeast Brazil	NEB	45°W -35°W	15°S - 2.5°S
Southern Amazonia	SAz	65°W - 50°W	17.5°S - 10°S
Southern Brazil	SB	60°W - 45°W	35°S - 22.5°S
East Pacific	EP	150°W - 90°W	5°S - 5°N
West Pacific	WP	120°E - 180°E	5°S - 5°N
South Atlantic	SA	40°W - 5°E	25°S - 5°S
North Atlantic	NA	75°W - 30°W	15°N - 35°N

Source: Table generated for the report by Cox and Jupp 2009.

Figure A2.1. Rainfall-Derived Model Probability Density Functions for the Eastern Amazonia Region (EA)

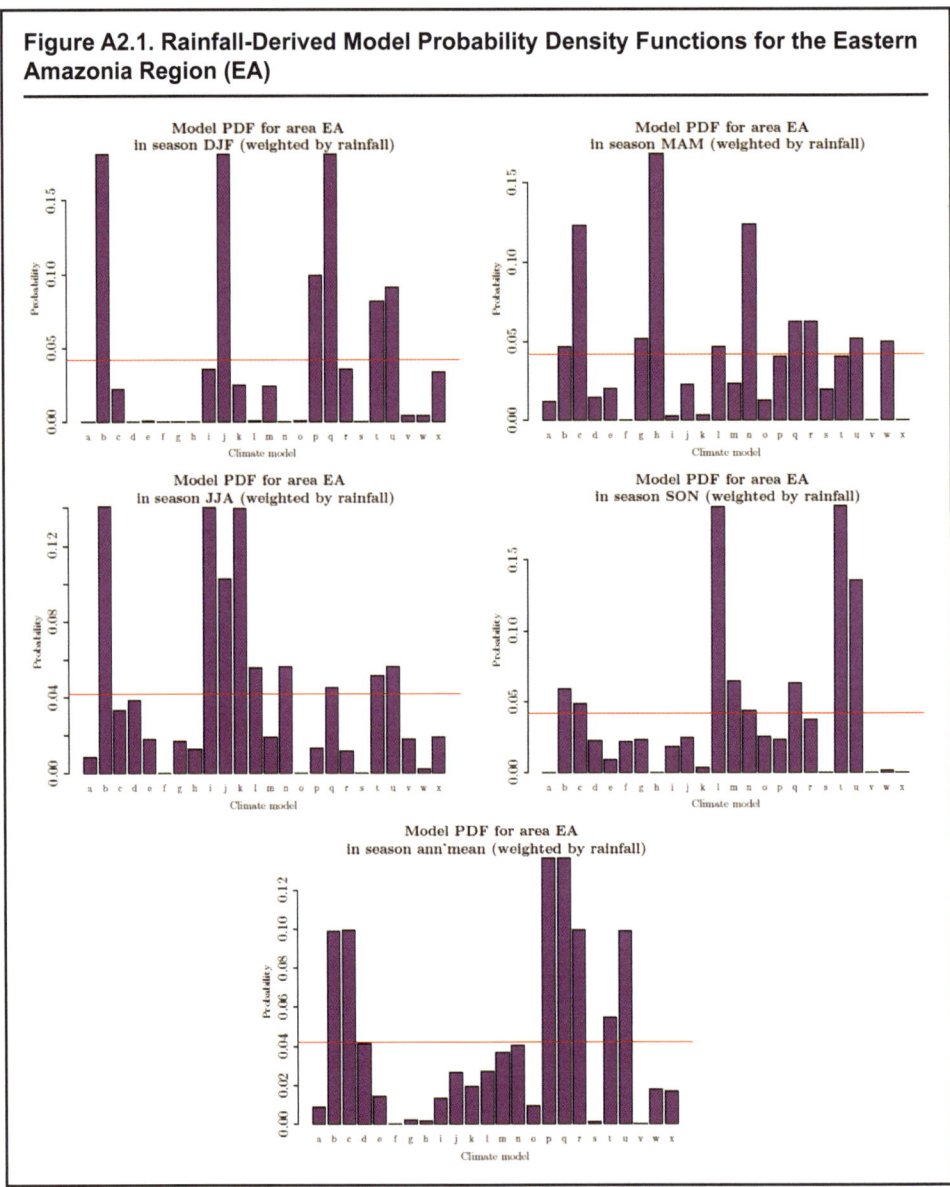

Source: Figure generated for the report by Cox and Jupp 2009.
Note: Models are labeled as in table A2.1.

Figure A2.2. Rainfall-Derived Model Probability Density Functions for the Northeast Brazil Region (NEB)

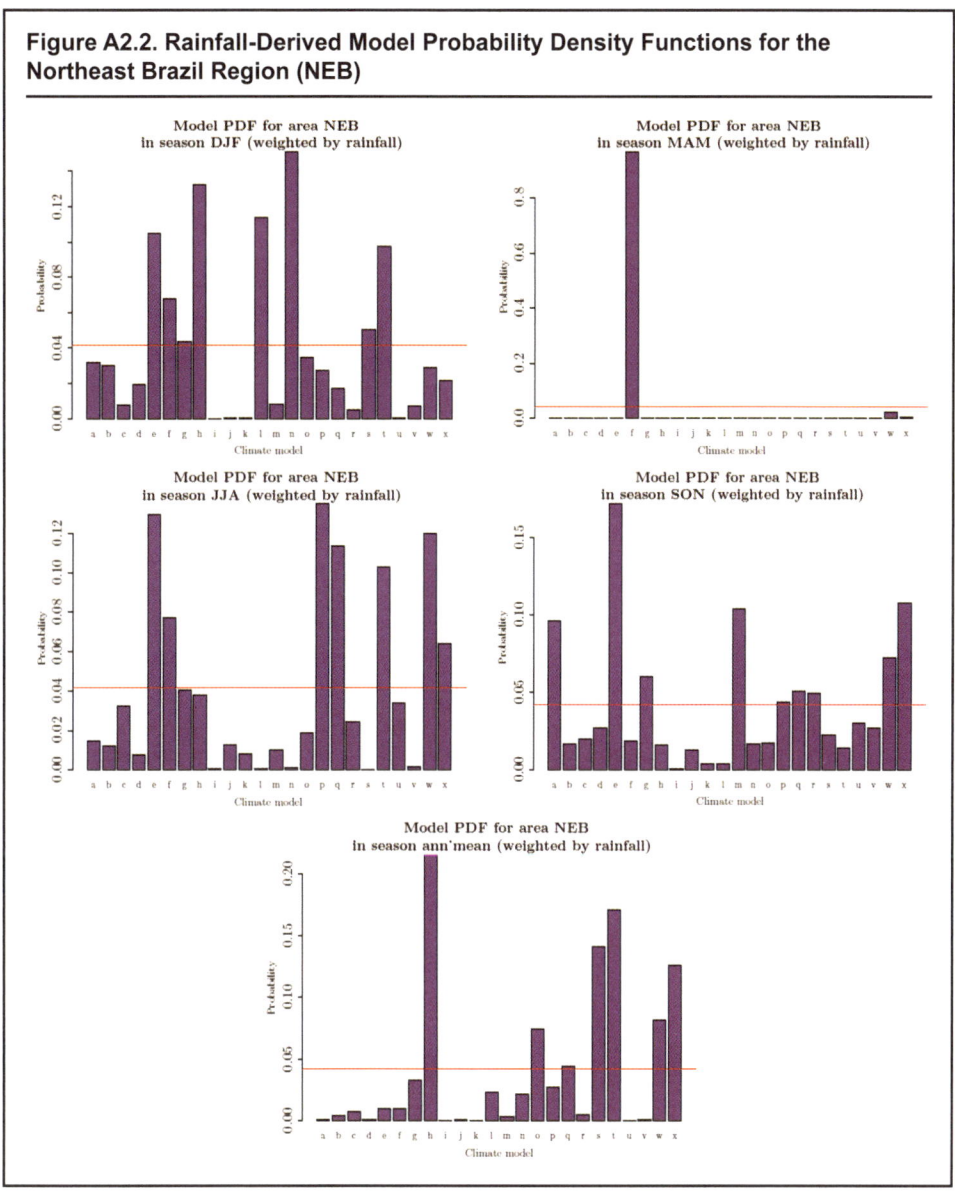

Source: Figure generated for the report by Cox and Jupp 2009.
Note: Models are labeled as in table A2.1.

Figure A2.3. Rainfall-Derived Model Probability Density Functions for the Northwest Amazonia Region (NWA)

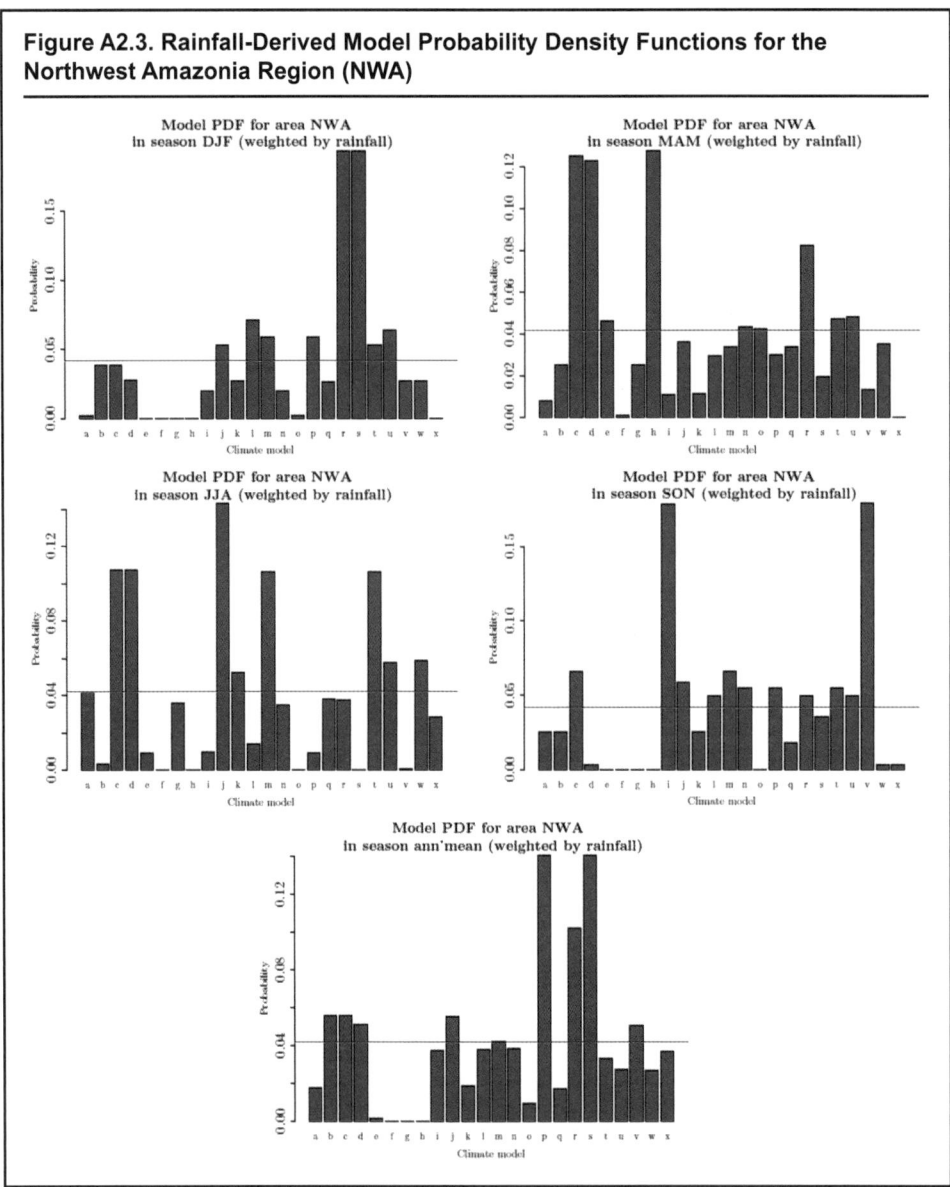

Source: Figure generated for the report by Cox and Jupp 2009.
Note: Models are labeled as in table A2.1.

Figure A2.4. Rainfall-Derived Model Probability Density Functions for the Southern Amazonia Region (SAz)

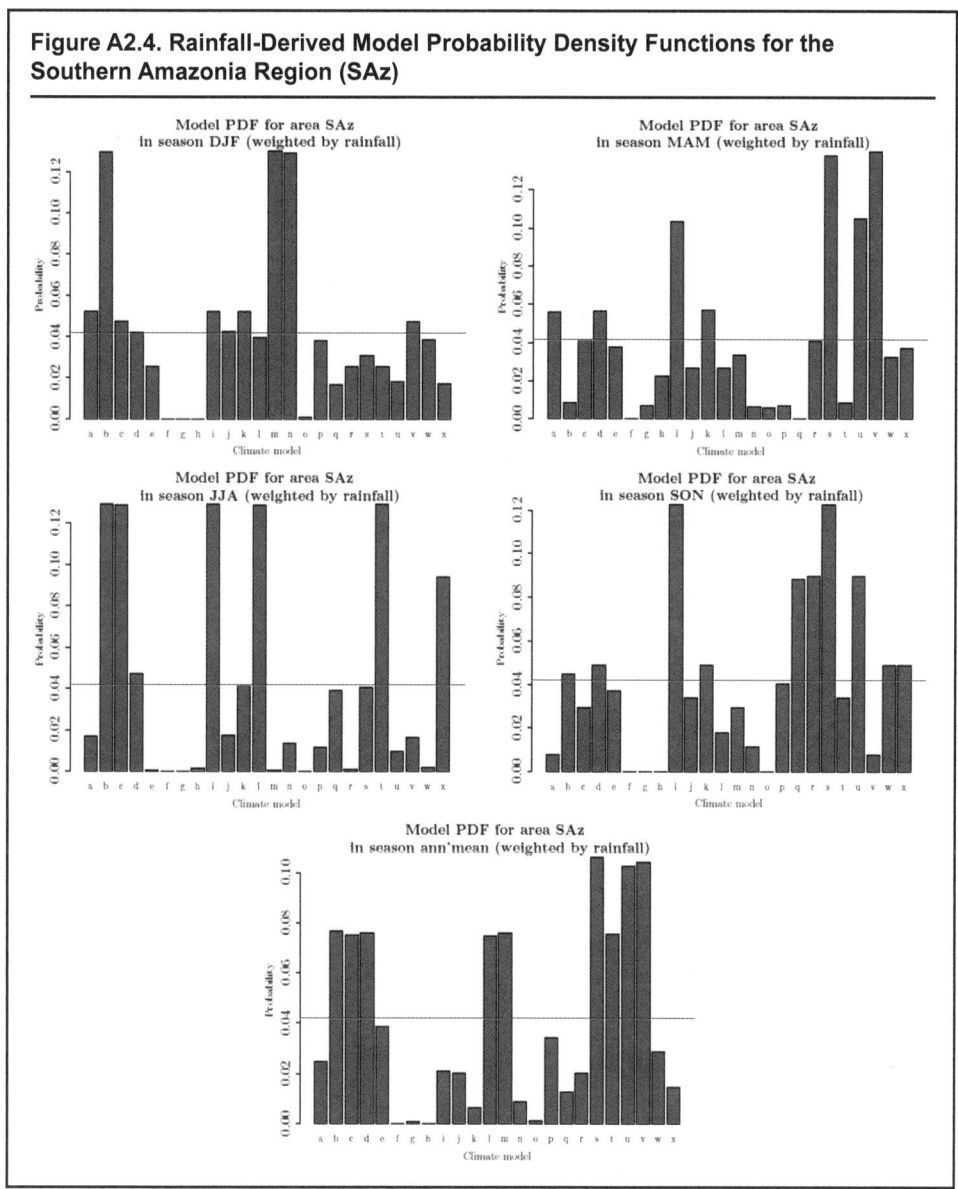

Source: Figure generated for the report by Cox and Jupp 2009.
Note: Models are labeled as in table A2.1.

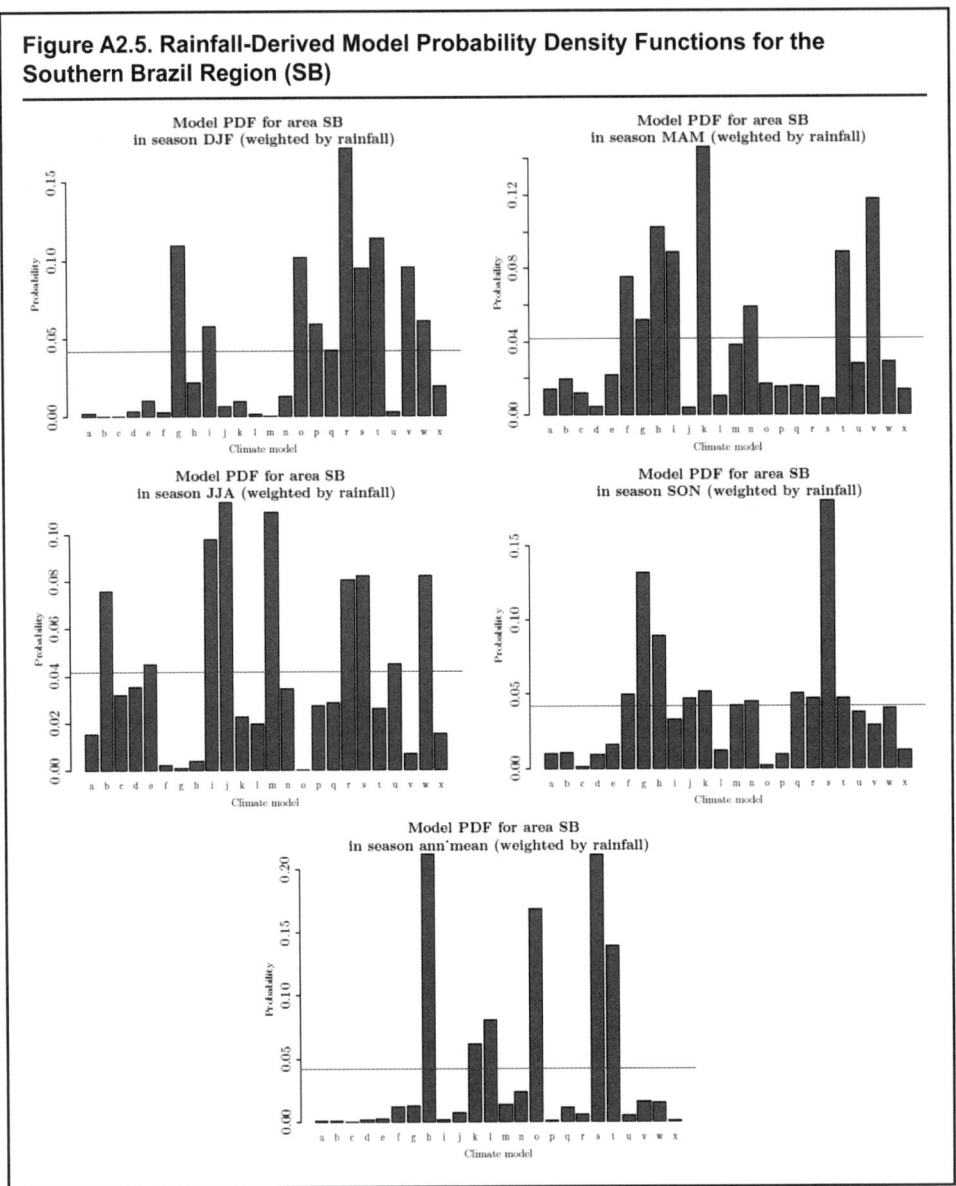

Figure A2.5. Rainfall-Derived Model Probability Density Functions for the Southern Brazil Region (SB)

Source: Figure generated for the report by Cox and Jupp 2009.
Note: Models are labeled as in table A2.1.

Figure A2.6. Sea Surface Temperature (ANSG)-Derived Model Probability Density Functions

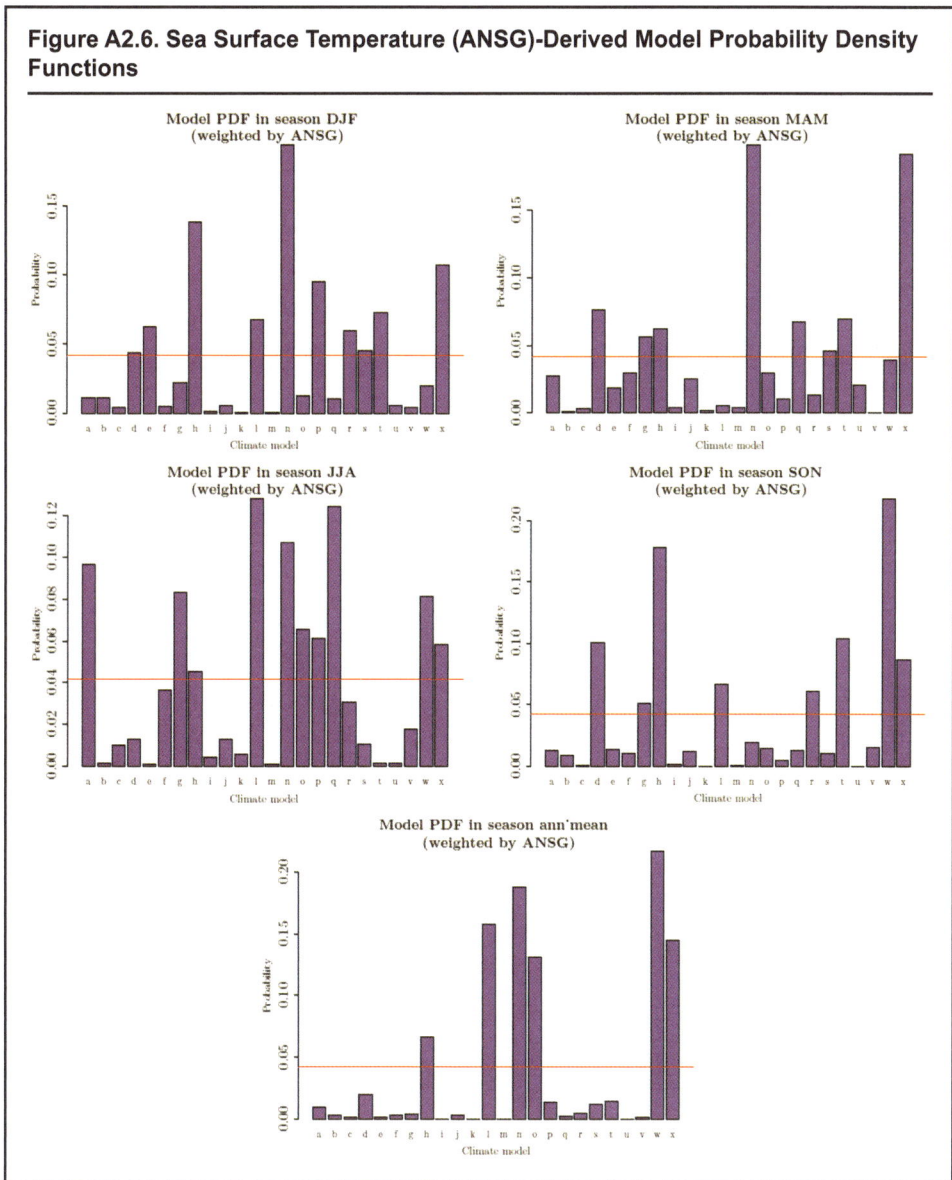

Source: Figure generated for the report by Cox and Jupp 2009.
Note: Models are labeled as in table A2.1.

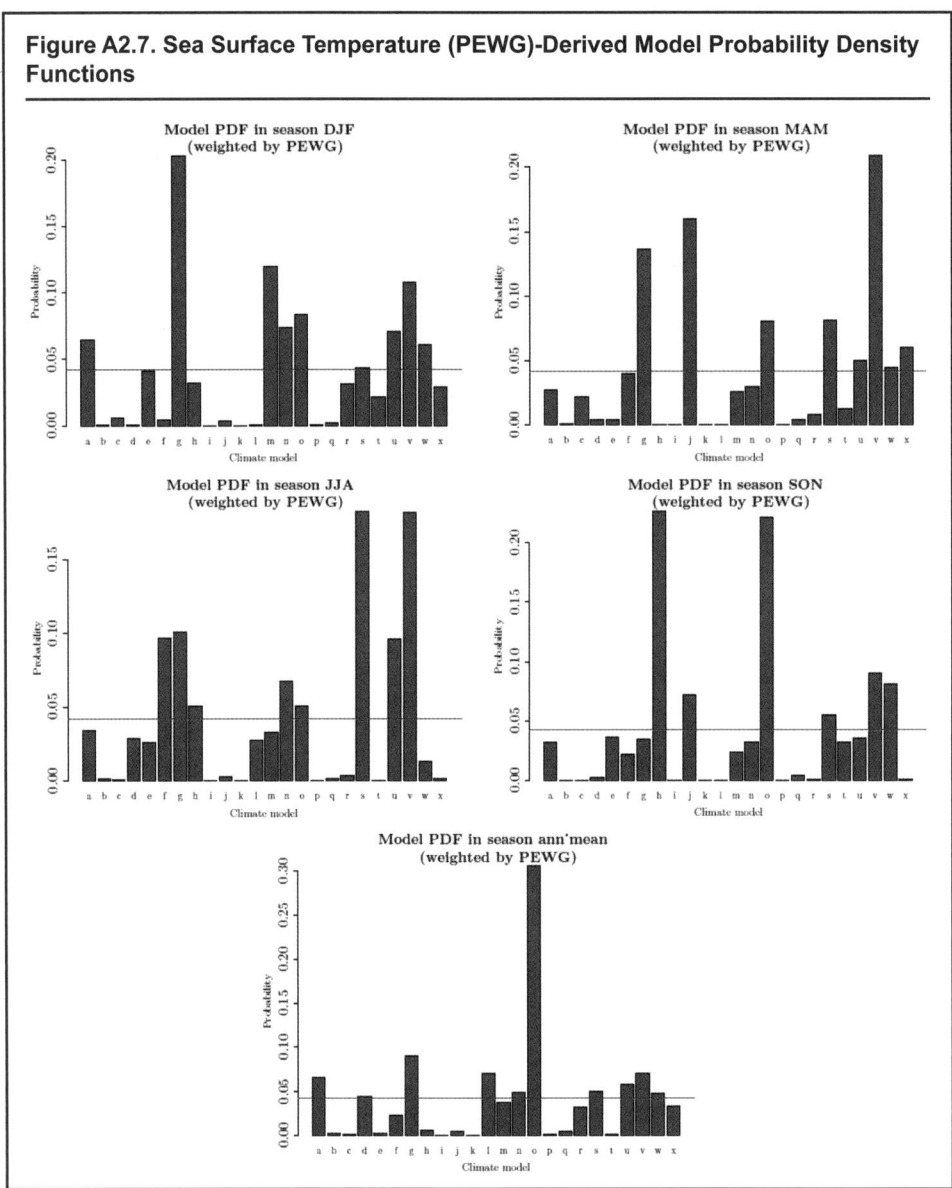

Figure A2.7. Sea Surface Temperature (PEWG)-Derived Model Probability Density Functions

Source: Figure generated for the report by Cox and Jupp 2009.
Note: Models are labeled as in table A2.1.

Figure A2.8. Sea Surface Temperature (ANSG and PEWG)-Derived Model Probability Density Functions

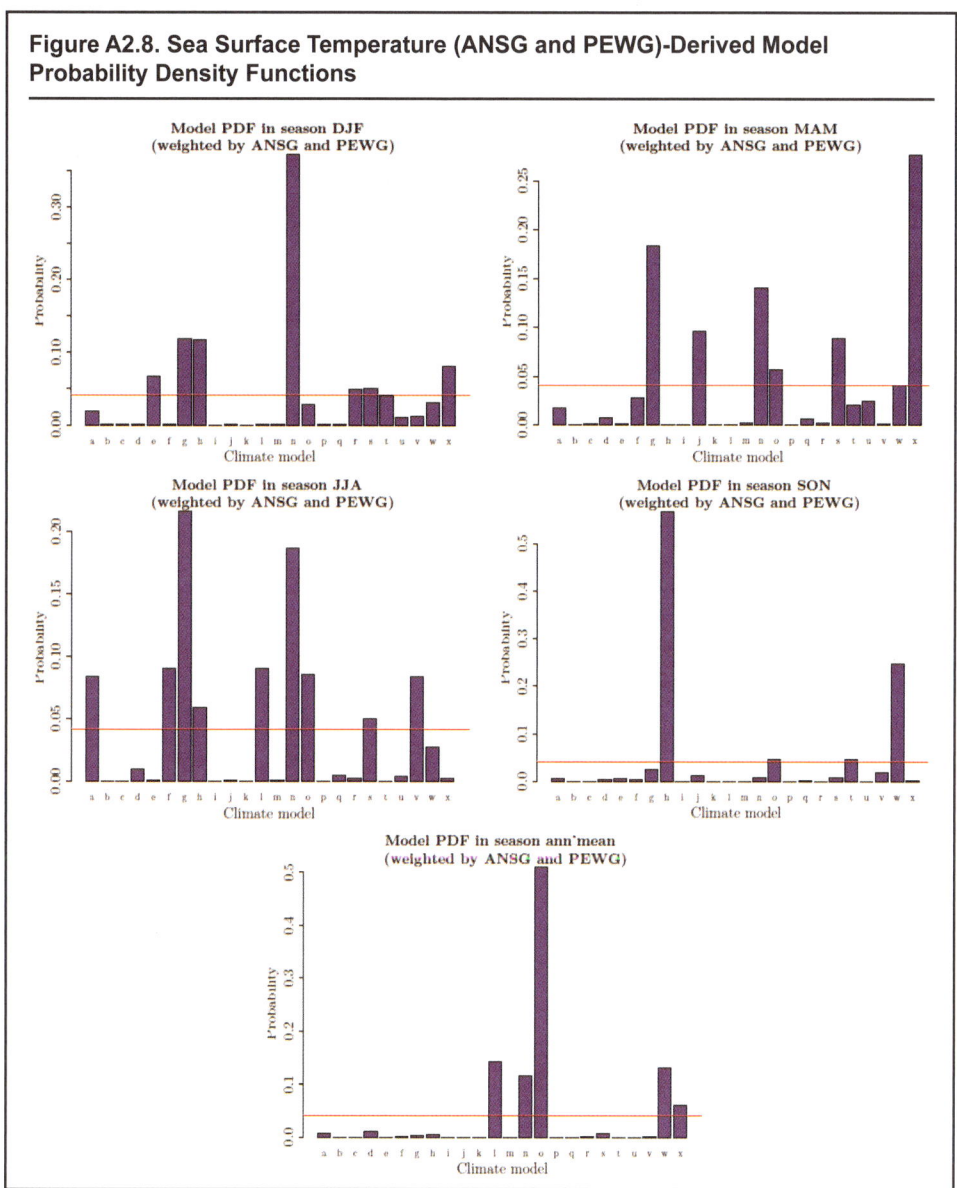

Source: Figure generated for the report by Cox and Jupp 2009.
Note: Models are labeled as in table A2.1.

Figure A2.9. Changes in Modeled Rainfall Cumulative Distribution Functions and Probability Density Functions for the Eastern Amazonia Region (EA) over the 21st Century

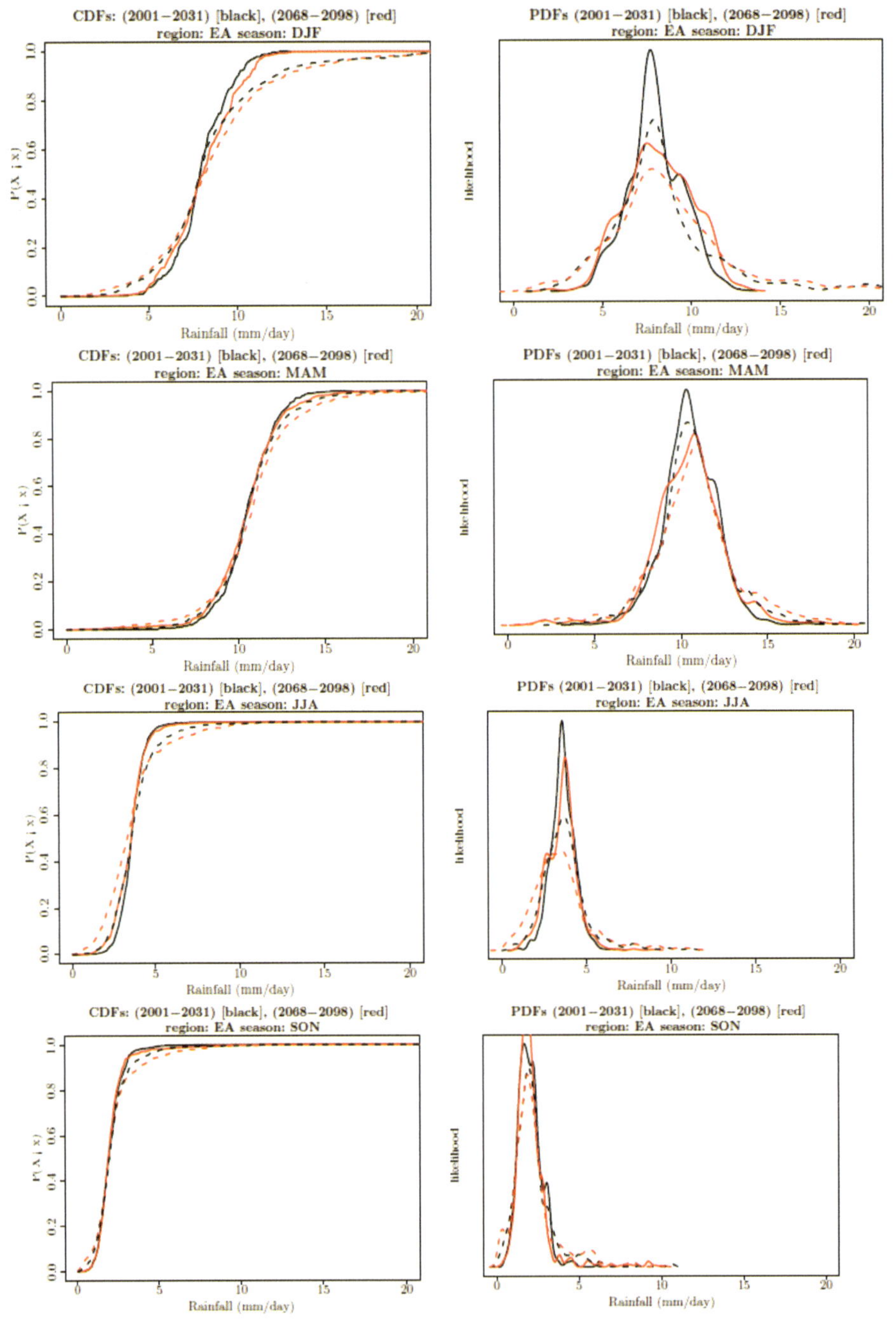

Source: Figure generated for the report by Cox and Jupp 2009.
Notes: Dotted lines: predictions with all models weighted equally. Solid lines: predictions with models weighted differentially according to the appropriate seasonal model probability density functions in figure A2.1.

Figure A2.10. Changes in Modeled Rainfall Cumulative Distribution Functions and Probability Density Functions for the Northeast Brazil Region (NEB) over the 21st Century

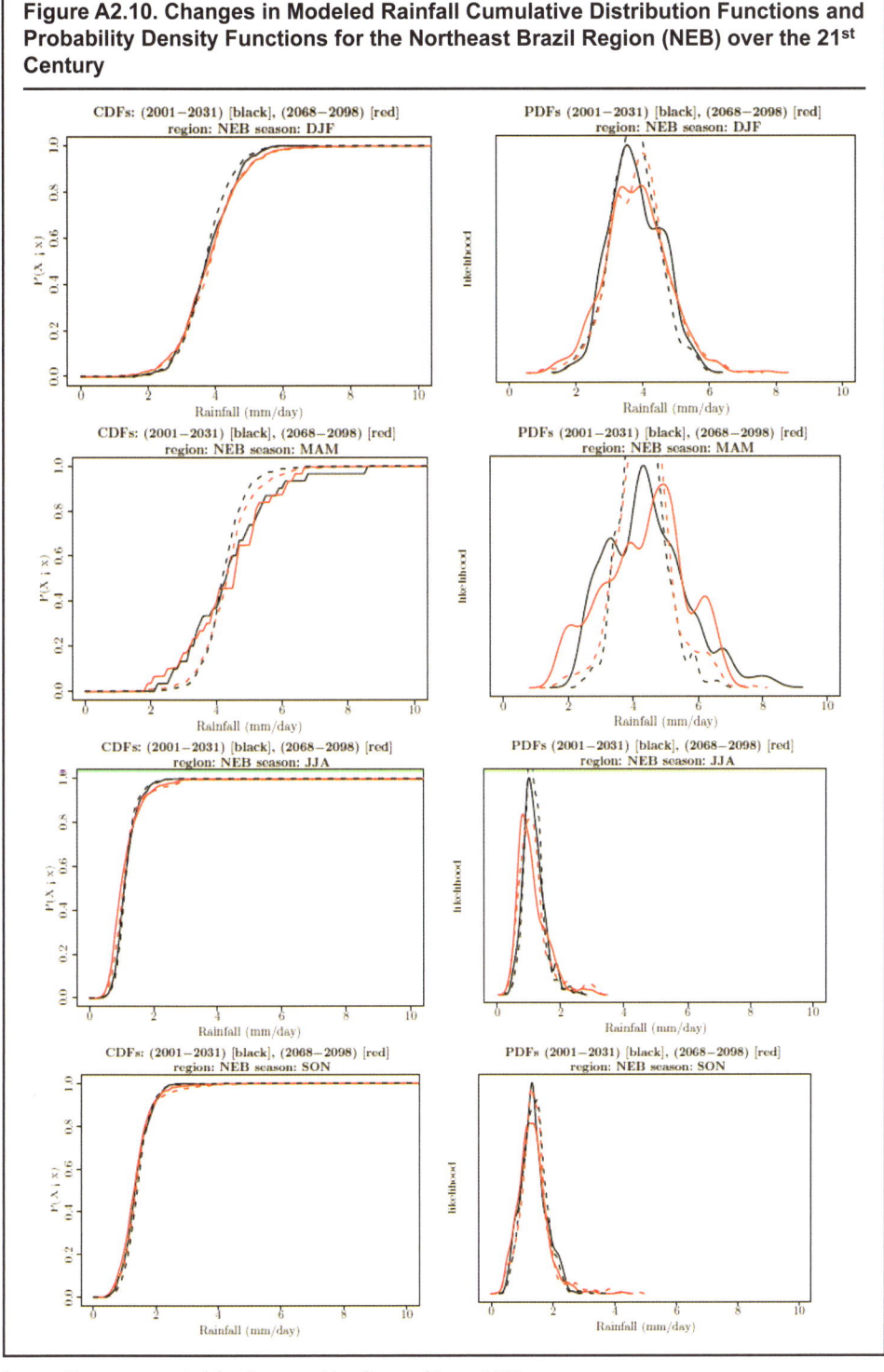

Source: Figure generated for the report by Cox and Jupp 2009.

Notes: Dotted lines: predictions with all models weighted equally. Solid lines: predictions with models weighted differentially according to the appropriate seasonal model probability density functions in figure A2.2.

Figure A2.11. Changes in Modeled Rainfall Cumulative Distribution Functions and Probability Density Functions for the Northwest Amazonia Region (NWA) over the 21st Century

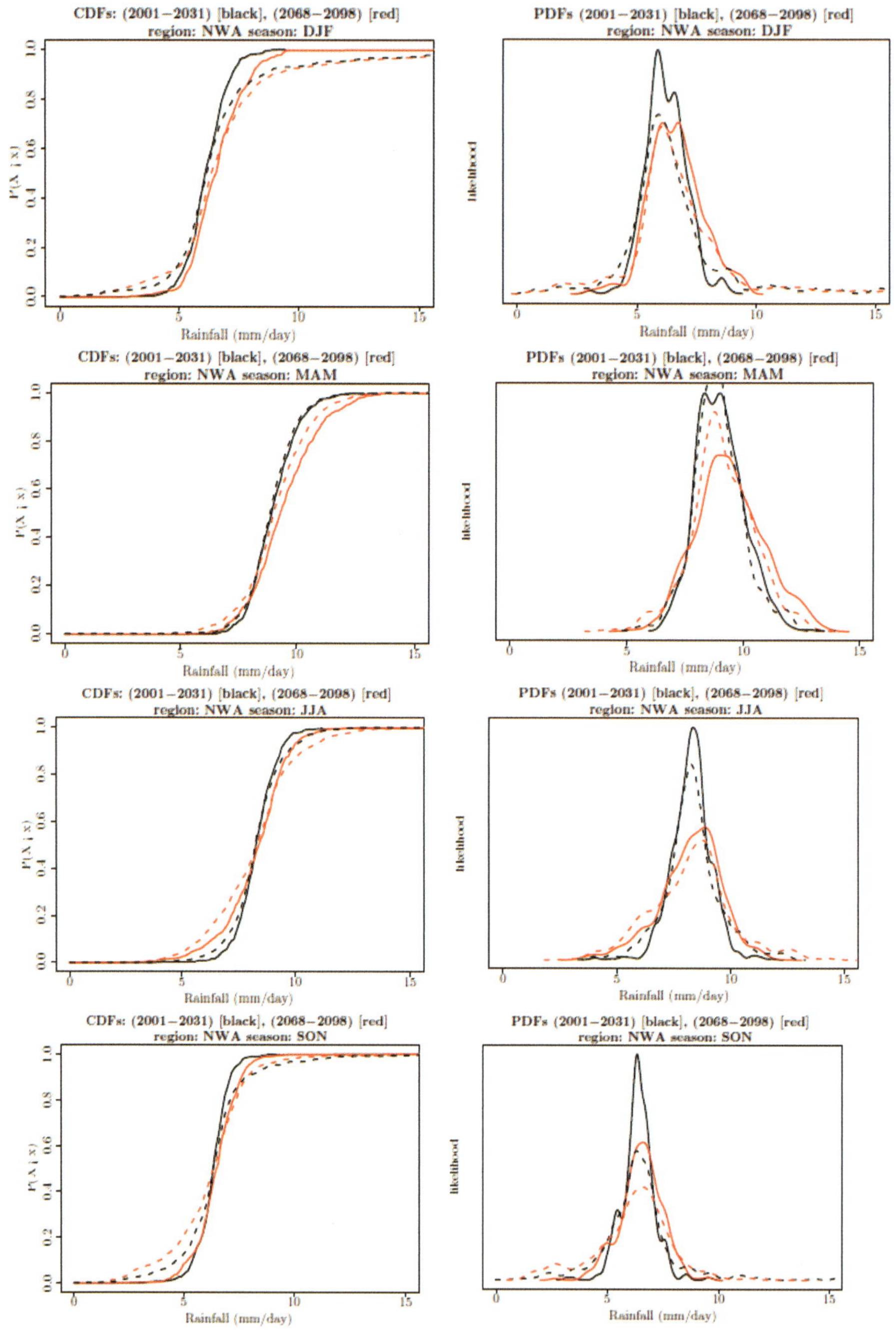

Source: Figure generated for the report by Cox and Jupp 2009.
Notes: Dotted lines: predictions with all models weighted equally. Solid lines: predictions with models weighted differentially according to the appropriate seasonal model probability density functions in figure A2.3.

Figure A2.12. Changes in Modeled Rainfall Cumulative Distribution Functions and Probability Density Functions for the Southern Amazonia Region (SAz) over the 21st Century

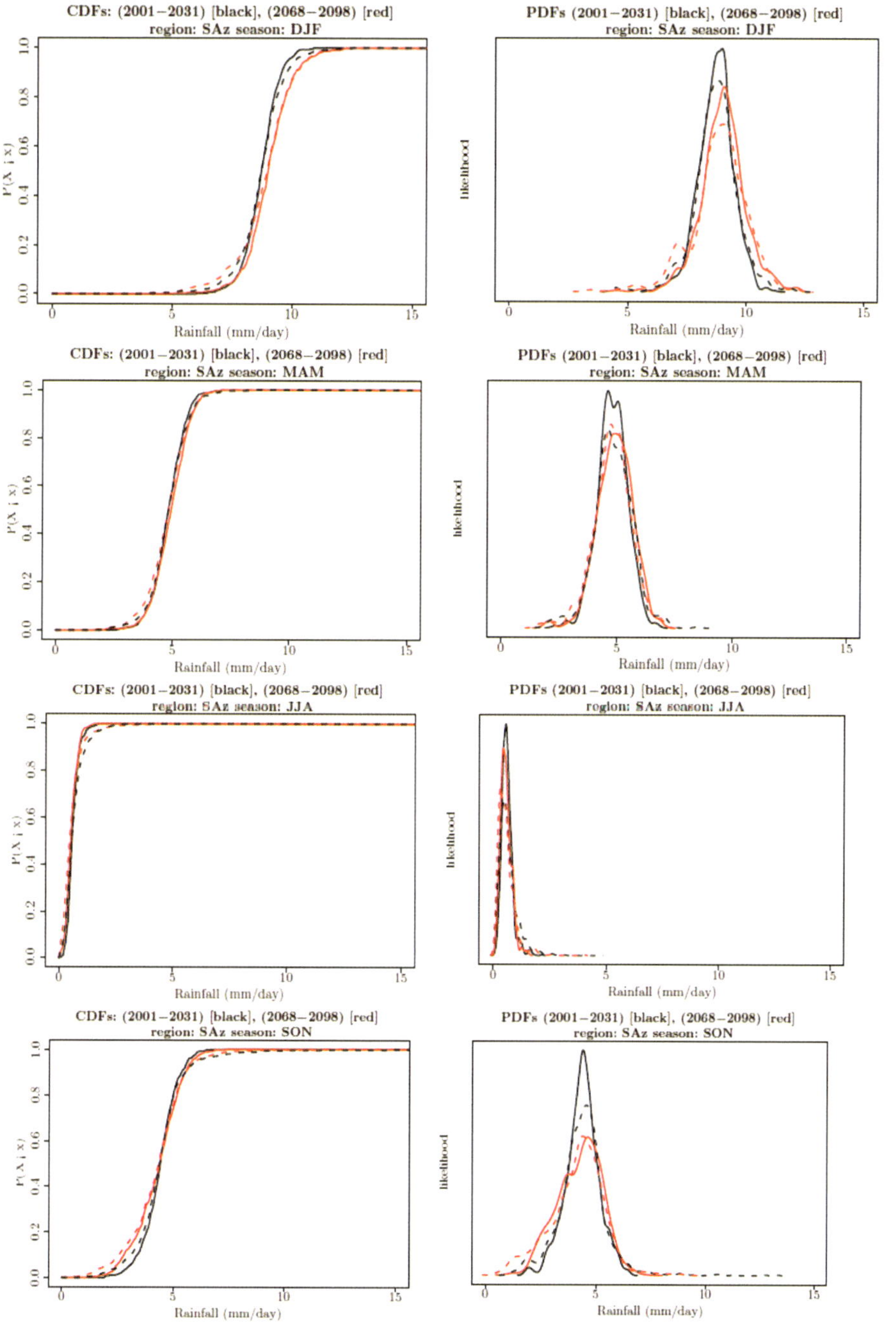

Source: Figure generated for the report by Cox and Jupp 2009.
Notes: Dotted lines: predictions with all models weighted equally. Solid lines: predictions with models weighted differentially according to the appropriate seasonal model probability density functions in figure A2.4.

Figure A2.13. Changes in Modeled Rainfall Cumulative Distribution Functions and Probability Density Functions for the Southern Brazil Region (SB) over the 21st Century

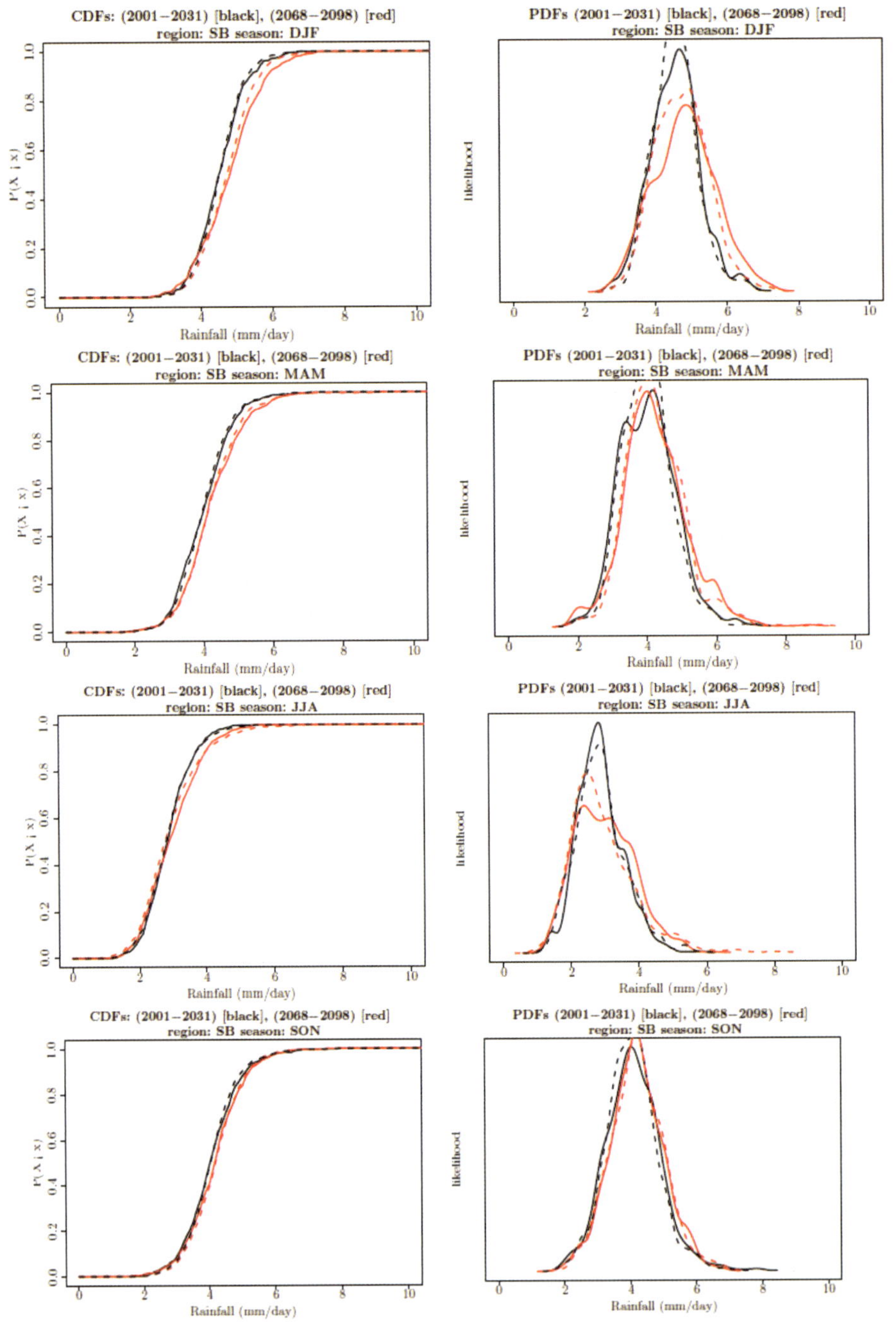

Source: Figure generated for the report by Cox and Jupp 2009.
Notes: Dotted lines: predictions with all models weighted equally. Solid lines: predictions with models weighted differentially according to the appropriate seasonal model probability density functions in figure A2.5.

Appendix C. Blue-Ribbon Panel Members—Short Biographies

Thomas E. Lovejoy (Chair)
President, The H. John Heinz Center for Science, Economics, and the Environment
Tropical biologist, conservation biologist, Ph.D. Biology, Yale University.
Lovejoy has worked in the Amazon since 1965. Since May 2002 he has been President of The Heinz Center. Before that, he was the World Bank's Chief Biodiversity Advisor and Lead Specialist for Environment for Latin America and the Caribbean and Senior Advisor to the President of the United Nations Foundation. Lovejoy has been Assistant Secretary and Counselor to the Secretary at the Smithsonian Institution, Science Advisor to the Secretary of the Interior, and Executive Vice President of the World Wildlife Fund–U.S.

Lawrence E. Buja
Project Manager, Climate Change Research Section/Climate and Global Dynamics Division, National Center for Atmospheric Research
Climate modeler, Ph.D. Atmospheric Sciences, University of Utah.
Buja is the scientific project manager of the Climate Change and Prediction group in the Climate Change Research Section at the National Center for Atmospheric Research (NCAR), Boulder, Colorado. Most recently, this group carried out the CCSM climate model simulations that made up the joint US NSF/DOE submission to the Intergovernmental Panel on Climate Change (IPCC) Fourth Assessment Report.

David Lawrence
Project Scientist, Climate Change Research Section/Climate and Global Dynamics Division, National Center for Atmospheric Research
Climate modeler, Ph.D. Atmospheric and Oceanic Science, University of Colorado.
Lawrence has used and contributed to the development of global climate models since 2001. For the last two years, he has served as co-chair of the NCAR Community Climate System Model Land Model Working Group. His research interests center around the role of land-surface processes and land-atmosphere interactions in climate change.

Michael T. Coe
Associate Scientist, Woods Hole Research Center
Atmospheric and Oceanic Scientist, Ph.D., University of Wisconsin-Madison.
Coe is an earth system scientist who is particularly interested in the causes and consequences of water resource variability. He uses data and earth system computer models to study how climate variability interacts with human land and water management practices to cause changes in water quality and quantity. He is currently participating in projects in the Amazon and Mississippi River basins as well as the semiarid regions of northern Africa. Coe previously spent seven years as a scientist at the Center for Sustainability and the Global Environment, University of Wisconsin-Madison and has been a visiting scientist at Lund University, Sweden, and the Max Planck Institute for Biogeochemistry, Jena, Germany.

Earl Saxon
Independent Consultant
Saxon's Ph.D. on the impacts of global warming and rising sea levels (Cambridge University, UK) was completed in 1975. He is an independent consultant on climate-adaptive tropical forest management strategies and related international policy negotiations. Previous appointments include Climate Policy Coordinator at the International Union for the Conservation of Nature, Lead Climate Scientist at The Nature Conservancy, Heritage Conservation Manager at the Wet Tropics of Queensland World Heritage Authority and Bullard Fellow at Harvard University. Saxon has conducted ecosystem research and forest conservation planning in Australia, Bolivia, China, Indonesia, Myanmar, Papua New Guinea, and the Philippines.

Ben Braga
Professor of Civil and Environmental Engineering, University of São Paulo (on leave of absence), Director of the National Water Agency of Brazil (ANA) and President of the International Hydrologic Program of UNESCO
Ph.D. in Water Resources Management, Stanford University. Research interests include integrated water resources management, conflict resolution, and multi-objective decision making. Braga worked in the Amazon Basin in the 1990s developing studies of the impact of deforestation on the hydrologic regime of small river basins. Currently as Director of ANA, he is in charge of coordinating the implementation of the National Water Resources Management System of Brazil.

Appendix D. Scientific Team of the Amazon Dieback Risk Assessment Task

Meteorological Research Institute (MRI), Japan
Akio Kitoh (Lead Author)
Shoji Kusunoki
Hiroki Kondo

University of Exeter, UK
Peter Cox (Lead Author)
Tim Jupp

Potsdam Institute for Climate Impact Research (PIK), Germany
Anja Rammig (Lead Author)
Wolfgang Cramer
Wolfgang Lucht
Kirsten Thonicke
Ursula Heyder

CCST/INPE, Brazil
Gilvan Sampaio (Lead Author)
Carlos Nobre
José Marengo
Lincoln Alves
José Fernando Pesquero
Celso Von Randow
Luis Fernando Salazar
Manoel Ferreira Cardoso

Earth 3000—GEXI, Germany and UK
Maritta Koch-Weser

RESTEC, Japan
Yukio Haruyama
Nobuhiro Tomiyama

World Bank Task Team
Walter Vergara (TTL)
Sebastian M. Scholz
Alejandro Deeb
Natsuko Toba
Adriana Valencia
Alonso Zarzar
Keiko Ashida

References

Alencar A., D. Nepstad and M.C.V. Diaz. 2006. "Forest understorey fire in the Brazilian Amazon in ENSO and non-ENSO years: area burned and committed carbon emissions." *Earth Interactions* 10: 1-17

Ainsworth, E.A., and S.P. Long. 2005. "What have we learned from 15 years of free-air CO2 enrichment (FACE)? A meta-analysis of the responses of photosynthesis, canopy properties and plant production to rising CO_2." *New Phytol.* 165: 351-372.

Bondeau A., P.C. Smith, S. Zaehle, S. Schaphoff, W. Lucht, W. Cramer, D. Gerten, H. Lotze-Campen, C. Müller, M. Reichstein and B. Smith. 2007. "Modelling the role of agriculture for the 20th century global terrestrial carbon balance." *Global Change Biology* 13: 679-706.

Calfapietra, C., B. Gielen, A.N.J. Galemma, M. Lukac, P. De Angelis, M.C. Moscatelli, R. Ceulemans, and G. Scarascia-Mugnozza. 2003. "Free-air CO_2 enrichment (FACE) enhances biomass production in a short-rotation poplar plantation." *Tree Phys.* 23: 805-814.

Cardoso, M. F., C. A. Nobre, D. M. Lapola, M. D. Oyama, and G. Sampaio, 2008. "Long-term potential for fires in estimates of the occurrence of savannas in the tropics." *Global Ecol. Biogeogr.* 17(2): 222-235.

Cardoso, M., C. Nobre, G. Sampaio, M. Hirota, D. Valeriano and G. Camera. 2009. "Long-term potential for tropical-forest degradation due to deforestation and fires in the Brazilian Amazon." *Biologia* 64(3): 433-437.

Cavalcanti, I.F.A., J.A. Marengo, P. Satyamurti, C.A. Nobre, I. Trosnikov, J.P. Bonatti, A.O. Manzi, T. Tarasova, L.P. Pezzi, C. D'Almeida, G. Sampaio, C.C. Castro, M.B. Sanches, M.B., and H. Camargo. 2002. "Global Climatological Features in a Simulation Using the CPTECCOLA AGCM." *Journal of Climate* 15(21): 2965-2988.

Cochrane M.A., and W.F. Laurance. 2008. "Synergisms among fire, land use, and climate change in the Amazon." *Ambio* 37: 522-527.

Collatz, G.J., J.T. Ball, C. Grivet, and J.A. Berry. 1991. "Physiological and environmental regulation of stomatal conductance, photosynthesis and transpiration: a model that includes a laminar boundary layer." *Agr. Forest Meteorol.* 54(2-4): 107-136.

Condit, R., S.P. Hubbell, and R.B. Foster. 1995. "Mortality rates of 205 neotropical tree and shrub species and the impact of a severe drought." *Ecological Monographs* 65: 419-439.

Condit R, S. Aguilar, A. Hernandez, R. Perez, S. Lao, G. Angehr, S.P. Hubbell, and R.B. Foster. 2004. "Tropical forest dynamics across a rainfall gradient and the impact of an El Nino dry season." *Journal of Tropical Ecology* 20: 51-72.

Cowling, S.A., and Y. Shin. 2006. "Simulated ecosystem threshold responses to co-varying temperature, precipitation and atmospheric CO_2 within a region of Amazonia." *Global Ecology & Biogeography* 15: 553-566.

Cox, P.M. 2001. "Description of the TRIFFID Dynamic Global Vegetation Model." UK, Hadley Centre, Met Office: 16.

Cox, P.M., and T.E. Jupp. 2008. "The sensitivity of Amazonian rainfall to SST indices on interannual timescales, and predicted rainfall change in the 21st century." World Bank PROJECT P109761, Discussion Paper 1.

Cox, P.M., and T.E. Jupp. 2009. "Development of probability density functions (PDFs) for future Amazonian rainfall." Report produced by University of Exeter for the World Bank as a background for this report.

Cox, P.M., R.A. Betts, M. Collins, P.P. Harris, C. Huntingford, and C. Jones. 2004. "Amazonian forest dieback under climate-carbon cycle projections for the 21st century." *Theor. Appl. Climatol.* 78: 137-156.

Cox, P.M., P.P. Harris, C. Huntingford, R.A. Betts, M. Collins, C.D. Jones, T.E. Jupp, J.A. Marengo, and C.A. Nobre. 2008. "Increasing risk of Amazonian drought due to decreasing aerosol pollution." *Nature* (453): 212-216.

Cramer, W., A. Bondeau, F.I. Woodward, I.C. Prentice, R.A. Betts, V. Brovkin, P.M. Cox, V. Fisher, J.A. Foley, A.D. Friend, C. Kucharik, M.R. Lomas, N. Ramankutty, S. Sitch, B. Smith, A. White, and C. Young-Molling. 2001. "Global response of terrestrial ecosystem structure and function to CO_2 and climate change: results from six dynamic global vegetation models." *Global Change Biology* 7: 357-373.

Curtis, P.S., and X.S. Wang. 1998. "A meta-analysis of elevated CO_2 effects on woody plant mass, form, and physiology." *Oecologia* 113: 299-313.

Dirzo, R., and P.H. Raven. 2003. "Global state of biodiversity and loss." *Annual Review of Environment and Resources* 28: 137-167.

Dorman, J.L., and P.J. Sellers. 1989. "A global climatology of albedo, roughness length and stomatal resistance for atmospheric general circulation models as represented by the Simple Biosphere model (SiB)." *J. Appl. Meteor.* 28: 833-855.

Essery, R.L.H., M.J. Best, and P.M. Cox. 2001. "MOSES 2.2 Technical Documentation." UK, Hadley Centre, Met Office: 30.

Farquhar, G.D., S. von Caemmerer, and J.A. Berry. 1980. "A biochemical model of photosynthesic CO_2 assimilation in leaves of C3 plants." *Planta* 149: 78-90.

Foley, J.A., I.C. Prentice, N. Ramankutty, S. Levis, D. Pollard, S. Sitch, and A. Haxeltine. 1996. "An integrated biosphere model of land surface processes, terrestrial carbon balance, and vegetation dynamics." *Global Biogeochemical Cycles* 10.

Folland, C.K., A.W. Colman, D.P. Rowell, and M.K. Davey. 2001. "Predictability of Northeast Brazil rainfall and real-time forecast skill 1987-1998." *Journal of Climate* 14: 1937-1958.

Fu, R., R.E. Dickinson, M.X. Chen, and H. Wang. 2001. "How do tropical sea surface temperatures influence the seasonal distribution of precipitation in the equatorial Amazon?" *Journal of Climate* 14: 4003-4026.

Gash, J.H.C., and C.A. Nobre. 1997. "Climatic effects of Amazonian deforestation: some results from ABRACOS." *Bulletin of the American Meteorological Society* 78(5): 823-830.

Gedney, N., P. M. Cox, R. A. Betts, O. Boucher, C. Huntingford, and P. A. Stott. 2006. "Detection of a direct carbon dioxide effect in continental river runoff records." *Nature* 439: 835-838.

Gerten, D., S. Schaphoff, U. Haberland, W. Lucht, and S. Sitch. 2004. "Terrestrial vegetation and water balance - hydrological evaluation of a dynamic global vegetation model." *Journal of Hydrology* 286: 249-270.

Gerten, D., W. Lucht, S. Schaphoff, W. Cramer, T. Hickler, and W. Wagner. 2005. "Hydrologic resilience of the terrestrial biosphere." *Geophysical Research Letters* 32: L21408.

Gifford, R.M. 2004. "The CO_2 fertilising effect—does it occur in the real world?" *New Phytol.* 163, 221-225.

Hickler, T., B. Smith, I.C. Prentice, K. Mjöfors, P. Miller, A. Arneth, and M. Sykes. 2008. "CO_2 fertilization in temperate FACE experiments not representative of boreal and tropical forests." *Global Change Biology* 14: 1531-1542.

Houghton, R.A., K.T. Lawrence, J.L. Hackler, and S. Brown. 2001. "The spatial distribution of forest biomass in the Brazilian Amazon: a comparison of estimates." *Global Change Biology* 7: 731-746.

Huffman, G.J., R.F. Adler, M.M. Morrissey, D.T. Bolvin, S. Curtis, R. Joyce, B. McGavock, and J. Susskind. 2001. "Global precipitation at one-degree daily resolution from multi-satellite observations." *J. Hydrometeorol.* 2, 36-50.

Hughes, J.K., P.J. Valdes, and R. Betts. 2006. "Dynamics of a globalscale vegetation model." *Ecological Modelling* 198: 452-462.

IBGE. 1988. Mapa de vegetacao do Brasil.

IPCC. 2007. *Climate Change (2007): The Physical Science Basis.* Contribution of Working Group I to the Fourth Assessment Report of the Intergovernmental Panel on Climate Change. Cambridge, United Kingdom and New York, NY, USA, Cambridge University Press. 996 pp.

Jablonski, L.M., X. Wang, and P.S. Curtis. 2002. "Plant reproduction under elevated CO_2 conditions: a meta-analysis of reports on 79 crop and wild species." *New Phytol.* 156: 9-26.

Kauffman, J.B., and C. Uhl. 1990. "Interactions of anthropogenic activities, fire, and rain forests in the Amazon basin." In *Fire in the Tropical Biota: Ecosystem Processes and Global Challenges.* Goldammer JG. Berlin, Springer Verlag: 117-134.

Kamiguchi, K., A. Kitoh, T. Uchiyama, R. Mizuta, and A. Noda. 2006. "Changes in precipitation-based extremes indices due to global warming projected by a global 20-km-mesh atmospheric model." *SOLA* 2: 64-67.

Kimball, B.A., K. Kobayashi, and M. Bindi. 2002. "Responses of agricultural crops to free-air CO_2 enrichment." *Adv Agron.* 77: 293-368.

Kitoh, A., S. Kusunoki, and KAKUSHIN Team3 Global Model Group. 2009. "Amazon climate in future simulated by 20-km and 60-km mesh atmospheric GCMs." Report produced by Meteorological Research Institute of Japan for the World Bank as a background for this report.

Kleidon, A., and M. Heimann, 2000. "Assessing the role of deep rooted vegetation in the climate system with model simulations: mechanism, comparison to observations and implications for Amazonian deforestation." *Climate Dynamics* 16: 183-199.

Korner, Ch. 2004. "Through enhanced tree dynamics carbon dioxide enrichment may cause tropical forests to lose carbon." *Philos Trans R Soc Lond Ser B-Biol Sci* 359:493-498.

Korner, Ch. 2006. "Plant CO_2 responses: an issue of definition, time and resource supply." *New Phytol* 172:393-411.

Korner, Ch., R. Asshoff, O. Bignucolo, S. Hattenschwiler, S.G. Keel, S. Pelaez-Riedl, S. Pepin, R.T.W. Siegwolf, and G. Zotz. 2005. "Carbon flux and growth in mature deciduous forest trees exposed to elevated CO_2." *Science* 309: 1360-1362.

Korner, Ch, J. Morgan, and R. Norby. 2007. "CO$_2$ Fertilisation: When, where, how much?" In Canadell J.G., Pataki D.E., and Pitelka L.F., eds. *Terrestrial Ecosystems in a Changing World Series: Global Change—The IGBP Series.* Springer, Berlin, p. 9-21.

Krinner, G., N. Viovy, N. de Noblet-Ducoudre, J. Ogee, J. Polcher, P. Friedlingstein, P. Ciais, S. Sitch, and I.C. Prentice. 2005. "A dynamic global vegetation model for studies of the coupled atmosphere biosphere system." *Global Biogeochemical Cycles* 19: GB1015.

Lapola, D.M., M.D. Oyama, C.A. Nobre, and G. Sampaio. 2008. "A new world natural vegetation map for global changes studies." *An. Acad. Bras. Cienc.* 80(2): 397-408.

Lapola, D.M., M.D. Oyama, and C.A. Nobre. 2009. "Exploring the range of climate-biome projections for tropical South America: the role of CO$_2$ fertilization and seasonality." *Global Biogeochemical Cycles*, Submitted.

Li, W., R. Fu, and R.E. Dickinson. 2006. "Rainfall and its seasonality over the Amazon in the 21st century as assessed by the coupled models for the IPCC AR4." *Journal of Geophysical Research* 111: D02111.

Liberloo, M., S.Y. Dillen, C. Calfapietra, S. Marinari, B.L. Zhi, P. De Angelis, and R. Ceulemans. 2005. "Elevated CO$_2$ concentration, fertilization and their interaction: growth stimulation in a short-rotation poplar coppice (EUROFACE)." *Tree Physiol.* 25: 179-189.

Liebmann, B., and J. Marengo. 2001. "Interannual variability of the rainy season and rainfall in the Brazilian Amazon basin." *Journal of Climate* 14: 4308-318.

Long, S.P., E.A. Ainsworth, A. Rogers, and D.R Ort. 2004. "Rising atmospheric carbon dioxide: Plants FACE the future." *Annual Rev. Plant Biol.* 55: 591-628.

Lucht, W., I.C. Prentice, R.B. Myneni, S. Sitch, P. Friedlingstein, W. Cramer, P. Bousquet, W. Buermann, and B. Smith. 2002. "Climatic control of the high-latitude vegetation greening trend and Pinatubo effect." *Science* 296: 1687-1689.

Malhi, Y., and J. Grace. 2000. "Tropical forests and atmospheric carbon dioxide." *Trends in Ecology & Evolution* 15: 332-337.

Malhi, Y., D. Wood, T.R. Baker, J. Wright, O.L. Phillips, T. Cochrane, P. Meir, J. Chave, S. Almeida, L. Arroyo, N. Higuchi, T.J. Killeen, S.G. Laurance, W.F. Laurance, S.L. Lewis, A. Monteagudo, D.A. Neill, P. Nunez Vargas, N.C.A. Pitman, C.A. Quesada, R. Salamao, J.N.M. Silva, A. Torres-Lezama, J. Terborgh, R. Vasquez Martinez, and B. Vinceti. 2006. "The regional variation of aboveground live biomass in old-growth Amazonian forests." *Global Change Biology* 12: 1107-1138.

Malhi, Y., J.T. Roberts, R.A. Betts, T.J. Killeen, W. Li, and C.A. Nobre. 2008. "Climate change, Deforestation, and the Fate of the Amazon." *Science* 319: 169-172.

Marengo, J.A. 2004. "Interdecadal variability and trends of rainfall across the Amazon basin." *Theoretical and Applied Climatology* 78: 79-96.

Meehl, G.A., and W.M. Washington. 1996. "El Nino-like climate change in a model with increased atmospheric CO$_2$ concentrations." *Nature* 382: 56-60.

Monserud, R.A., and R. Leemans. 1992. "Comparing global vegetation maps with the Kappa statistic." *Ecological Modelling* 62: 275-293.

Mizuta, R., K. Oouchi, H. Yoshimura, A. Noda, K. Katayama, S. Yukimoto, M. Hosaka, S. Kusunoki, H. Kawai, and M. Nakagawa. 2006. "20-km-mesh global climate simulations using JMA-GSM model –Mean climate states–." *J. Meteor. Soc. Japan* 84: 165-185.

Nakaegawa, T., and W. Vergara. 2006. "First Projection of Climatological Mean River Discharges in the Magdalena River Basin, Colombia, in a Changing Climate during the 21st Century." *Hydrological Research Letters* 4: 50–54 (2010). Published online in J-STAGE (www.jstage.jst.go.jp/browse/HRL).

Nepstad, D., C.R. De Carvalhos, E.A. Davidson, P.H. Jipp, P.A. Lefebvre, G.H. Negreiros, E.D. Da Silva, T.A. Stone, S.E. Trumbore, and S. Vieira. 1994. "The role of deep roots in the hydrological and carbon cycles of Amazonian forests and pastures." *Nature* 372: 666-669.

Nepstad, D., G. Carvalho, A.C. Barros, A. Alencar, J.P. Capobianco, J. Bishop, P. Moutinho, P. Lefebre, U. L. Silva Jr, and E. Prins. 2001. "Road paving, fire regime feedbacks, and the future of Amazon forests." *Forest Ecology and Management* 154: 395-407.

Nepstad, D.C., I.M. Tohver, D. Ray, P. Moutinho, and G. Cardinot. 2007. "Mortality of large trees and lianas following experimental drought in an Amazon forest." *Ecology* 88: 2259-2269.

New, M., M. Hulme, and P. Jones, 2000. "Representing 20th century space-time climate variability. Part II: Development of 1901–1996 monthly grids of terrestrial surface climate." *Journal of Climate* 13: 2217-2238.

Nobre, C.A., P.J. Sellers, and J. Shukla. 1991. "Amazonian deforestation and regional climate change." *Journal of Climate* 4: 957-988.

Nobre, P., and J. Shukla. 1996. "Variations of sea surface temperature, wind stress and rainfall over the tropical Atlantic and South America." *Journal of Climate* 9: 2464-2479.

Nohara, D., M. Hosaka, A. Kitoh, and T. Oki. 2006. "Impact of climate change on river runoff using multi-model ensemble." *J. Hydrometeorol.* 7: 1076-1089.

Norby, R.J., S.D. Wullschleger, C.A. Gunderson, D.W. Johnson, and R. Ceulemans. 1999. "Tree response to rising CO_2 in field experiments: implications for the future forest." *Plant, Cell and Environment* 22: 683-714.

Norby, R.J., J.D. Sholtis, C.A. Gunderson, and S.S. Jawdy. 2003. "Leaf dynamics of a deciduous forest canopy; no response to elevated CO_2." *Oecologia* 136: 574-584.

Norby, R.J., E.H. DeLucia, B. Gielen, C. Calfapietra, C.P. Giardina, J.S. King, J. Ledford, H.R. McCarthy, D.J.P. Moore, R. Ceulemans, P. De Angelis, A.C. Finzi, D.F. Karnosky, M.E. Kubiske, M. Lukac, K.S. Pregitzer, G.E. Scarascia-Mugnozza, W.H. Schlesinger, and R. Oren. 2005. "Forest response to elevated CO_2 is conserved across a broad range of productivity." *Proceedings of the National Academy of Sciences of the United States of America* 102: 18052-18056.

Nowak, R.S., D.S. Ellsworth, and S.D. Smith. 2004. "Tansley review: functional responses of plants to elevated atmospheric CO_2—Do photosynthetic and productivity data from FACE experiments support early predictions?" *New Phytol.* 162: 253-280.

Oyama, M.D., and C.A. Nobre. 2004. "A simple potencial vegetation model for coupling with the Simple Biosphere Model (SIB)." *Revista Brasileira de Meteorologia* 19(2): 203-216.

Phillips, O.L., S.L. Lewis, T.R. Baker, K-J. Chao, and N. Higuchi. 2008. "The changing Amazon forest." *Philosophical Transactions of the Royal Society Ser. B* 363: 1819-1827.

Phillips, O.L., L.E. Aragao, S.L. Lewis, J.B. Fisher, J. Lloyd, G. Lopez-Gonzalez, Y. Malhi, A. Monteagudo, J. Peacock, C.A. Quesada, G. van der Heijden, S. Almeida, I.

Amaral, L. Arroyo, G. Aymard, T.R. Baker, O. Banki, L. Blanc, D. Bonal, P. Brando, J. Chave, A.C. de Oliveira, N.D. Cardozo, C.I. Czimczik, T.R. Feldpausch, M.A. Freitas, E. Gloor, N. Higuchi, E. Jimenez, G. Lloyd, P. Meir, C. Mendoza, A. Morel, D.A. Neill, D. Nepstad, S. Patino, M.C. Penuela, A. Prieto, F. Ramirez, M. Schwarz, J. Silva, M. Silveira, A.S. Thomas, H.T. Steege, J. Stropp, R. Vasquez, P. Zelazowski, E. Alvarez Davila, S. Andelman, A. Andrade, K.J. Chao, T. Erwin, A. Di Fiore, C.E. Honorio, H. Keeling, T.J. Killeen, W.F. Laurance, A. Pena Cruz, N.C. Pitman, P. Nunez Vargas, H. Ramirez-Angulo, A. Rudas, R. Salamao, N. Silva, J. Terborgh, and A. Torres-Lezama. 2009. "Drought sensitivity of the Amazon rainforest." *Science* 323: 1344-7.

Prentice, I.C., A. Bondeau, W. Cramer, S.P. Harrison, T. Hickler, W. Lucht, S. Sitch, B. Smith, and M. Sykes. 2007. "Dynamic global vegetation modeling: Quantifying terrestrial ecosystem responses to large-scale environmental change." In *Terrestrial Ecosystems in a Changing World.* Canadell J.G., Pataki D.E., and Pitelka L.F. Berlin, Heidelberg, New York: Springer, 175-192.

Rammig, A., W. Cramer, W. Lucht, K. Thonicke, and U. Heyder. 2009. "Brazil: Risk analysis for Amazon dieback, 2009." Report produced by the Potsdam Institute for Climate Impact Research (PIK) for the World Bank as a background for this report.

Raschke, E., S. Bakan, and S. Kinne. 2006. "An assessment of radiation budget data provided by the ISCCP and GEWEX-SRB." *Geophys. Res. Lett.* 33: L07812.

Rayner, N.A., D.E. Parker, E.B. Horton, C.K. Folland, L.V. Alexander, D.P. Rowell, E.C. Kent, and A. Kaplan. 2003. "Global analyses of SST, sea ice and night marine air temperature since the late nineteenth century." *Journal of Geophysical Research* 108(D14): 10.1029/2002JD002670.

Saatchi, S.S., R.A. Houghton, R.C. Dos Santos Alvala, J.V. Soares, and Y. Yu. 2007. "Distribution of aboveground live biomass in the Amazon basin." *Global Change Biology* 13: 816-837.

Sahagian, D.L., and K. Hibbard. 1998. "GAIM 1993-1997, The first five years: setting the stage for synthesis." IGBP/GAIM Report Series, Report #6, 78p.

Salazar, L.F., C. Nobre, and M.D. Oyama. 2007. "Climate change consequences on the biome distribution in tropical South America." *Geophysical Research Letters* 34: L09708.

Sampaio, G., C. Nobre, M.H. Costa, P. Satyamurty, B.S. Soares-Filho, and M. Cardoso. 2007. "Regional climate change over eastern Amazonia caused by pasture and soybean cropland expansion." *Geophysical Research Letters* 34: L17709.

Sampaio, G., M. Ferreira Cardoso, and L.F. Salazar. 2009. "A Review of Amazon dieback and the research on tipping points: Simulations and new studies. Estimated tipping points for combined deforestation and climate impacts." Report produced by Brazilian Institute for Space Research-INPE for the World Bank as a background for this report.

Sellers, P.J., D.A. Randall, G.J. Collatz, J.A. Berry, C.B. Field, D.A. Dazlich, C. Zhang, G.D. Collelo, and L. Bounoua. 1996. "A revised land surface parameterization (SiB2) for Atmospheric GCMs. Part I: model formulation." *J. Climate* 9(4): 676-705.

Sitch, S., B. Smith, I.C. Prentice, A. Arneth, A. Bondeau, W. Cramer, J.O. Kaplans, S. Levis, W. Lucht, M.T. Sykes, K. Thonicke, and S. Venevsky. 2003. "Evaluation of ecosystem dynamics, plant geography and terrestrial carbon cycling in the LPJ dynamic global vegetation model." *Global Change Biology* 9: 161-185.

Sitch, S., C. Huntingford, N. Gedney, P.E. Levy, M. Lomass, S. Piao, R.A. Betts, P. Ciais, P.M. Cox, P. Friedlingstein, C.D. Jones, I.C. Prentice, and F.I. Woodward. 2008. "Evaluation of the terrestrial carbon cycle, future plant geography and climate-carbon cycle feedbacks using five Dynamic Global Vegetation Models (DGVMs)." *Global Change Biology* 14: 1-25.

Soares-Filho, B.S., D.C. Nepstad, L.M. Curran, G.C. Cerqueira, R.A. Garcia, C.A. Ramos, E. Voll, A. McDonald, P. Lefebvre, and P. Schlesinger. 2006. "Modelling conservation in the Amazon basin." *Nature* 440: 520-523.

Thonicke, K., A. Spessa, I.C. Prentice, S.P. Harrison, and C. Carmona-Morena, in review. "The influence of vegetation, fire spread and fire behaviour on global biomass burning and trace gas emissions." *Global Change Biology*.

Turner, D. P., W. D. Ritts, W. B. Cohen, S. T. Gower, S. W. Running, M. Zhao, M. H. Costa, A. A. Kirschbaum, J. M. Ham, S. R. Saleska, and D. E. Ahl. 2006. "Evaluation of MODIS NPP and GPP products across multiple biomes." *Remote Sens. Environ.* 102(3-4): 282-292.

Van Nieuwstadt, M.G.L., and D. Sheil. 2005. "Drought, fire and tree survival in a Borneo rain forest, East Kalimatan, Indonesia." *Journal of Ecology* 93: 191-201.

Vanhatalo, M., J. Back, and S. Huttunen. 2003. "Differential impacts of long-term (CO_2) and O_3 exposure on growth of northern conifer and deciduous tree species." *Trees-Struct. Funct.* 17: 211-220.

Williamson, G.B., S.G. Laurance, P.J.C. Oliveira, C. Gascon, T.E. Lovejoy, and L. Pohl. 2000. "Amazonia tree mortality during the 1997 El Nino drought." *Conservation Biology* 14: 1538-1542.

Willmott, C. J., C. M. Rowe, and Y. Mintz. 1985. "Climatology of the terrestrial seasonal water cycle." *Int. J. Climatology* 5(6): 589-606.

Willmott, C.J., and K. Matsuura. 1998. "Terrestrial air temperature and precipitation: monthly and annual climatologies." Online at: http://climate.geog.udel.edu/_climate/html pages/archive.html. University of Delaware, Newark.

Wittig, V.E., C.J. Bernacchi, X.-G. Zhu, C. Calfapietra, R. Ceulemans, P. Deangelis, B. Gielen, F. Miglietta, P.B. Morgan, and S.P. Long. 2005. "Gross primary production is stimulated for three *Populus* species grown under free-air CO_2 enrichment from planting through canopy closure." *Glob. Change Biol.* 11: 644-656.

Xie, P., and P.A. Arkin. 1997. "Global precipitation: A 17-year monthly analysis based on gauge observations, satellite estimates, and numerical model outputs." *Bull. Am. Meteorol. Soc.* 78: 2539-2558.

Xue, Y., P.J. Sellers, J.L. Kinter, and J.A. Shukla. 1991. "Simplified biosphere model for global climate studies." *Journal of Climate* 4: 345-364.